Magnetics Virtual Conference 2021

Online
19-20 January 2021

ISBN: 978-1-7138-3139-6

Printed from e-media with permission by:

Curran Associates, Inc.
57 Morehouse Lane
Red Hook, NY 12571

Some format issues inherent in the e-media version may also appear in this print version.

Printed with permission by Curran Associates, Inc. (2021)

TWST Events
622 3rd Ave, 34th FL
New York, NY 10017
USA

twstevents.com

Additional copies of this publication are available from:

Curran Associates, Inc.
57 Morehouse Lane
Red Hook, NY 12571 USA
Phone: 845-758-0400
Fax: 845-758-2633
Email: curran@proceedings.com
Web: www.proceedings.com

TABLE OF CONTENTS

MAGNETICS 2021

The Ideal Permanent Magnet – Myths and Realities.

Dr John Ormerod

January 20th, 2021

"The Nation That Controls Magnetism Will Control The Universe"

- Dick Tracy cartoon strip, created by Chester Gould.
- Circa early-1960's i.e., before rare earth magnets and the Chinese dominance of RE supply chain and magnet industry!

JOC LLC

Outline

- Introduction:
  - Me
  - Rare Earths
- The Ideal Permanent Magnet
  - Magnetics
  - Other Considerations
- Recent Magnet Material Developments
- Summary

JOC LLC

Introduction - Me

- BSc, MSC and PhD in Metallurgy from the University of Manchester (1972 – 1978).
- Magnetics career began for Philips (UK and Holland) – 1979 – 1990.
  - Developed and commercialized SmCo5, 2:17 and NdFeB magnets.
- Joined Arnold Magnetic Technologies (USA) responsible for soft and hard magnetic materials development and GM for permanent magnets (1990 – 2002).
- 2002 - 2014 President of Res Manufacturing in Milwaukee.
  - Metal stamping and value-added assemblies to the automotive market (Toyota, GM, Nissan).
  - Major supplier to Tesla Motors for Model S and Model X.
- Founded business and technology consultancy for magnetics and metals related industries in 2015 – JOC LLC (www.jocllc.com) assisted clients in all aspects of the PM supply chain.
- Provided expert testimony on key NdFeB patent challenges over the last few years.
- Advisory Board member for Bunting Group, Technology/Industry Advisor for Bunting, CMI and MP Materials.

JOC LLC

Introduction – Rare Earths

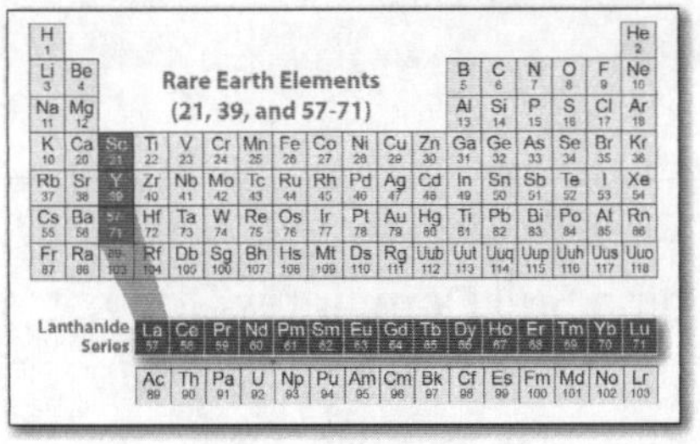

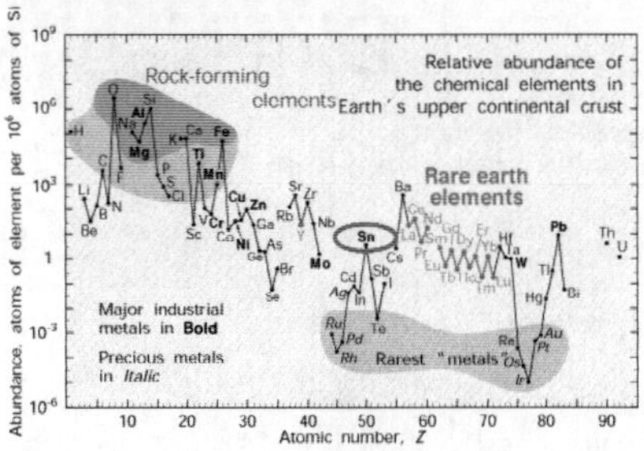

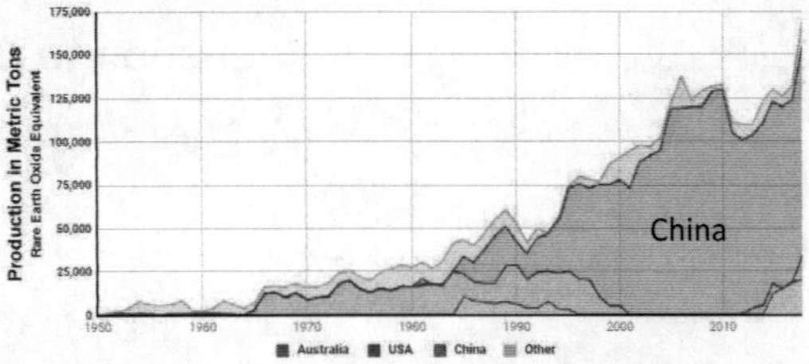

JOC LLC

China's REE dominance is (might be) diminishing

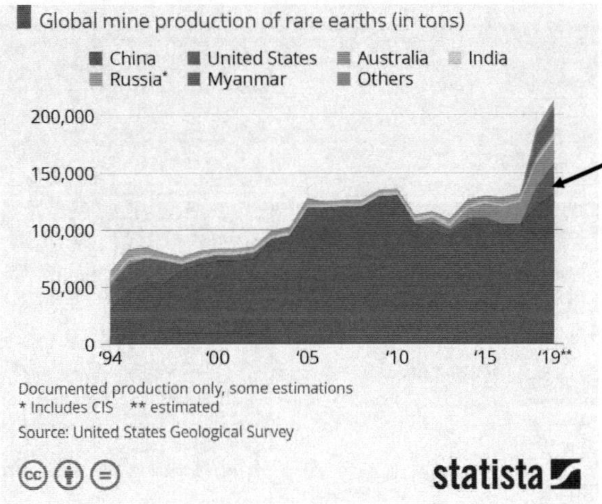

MP Materials' Mountain Pass back online.

JOC LLC

China's Dominance of Mining is Diminishing, but.........

On Top of the Mountain
China dominates global production of mined rare earths

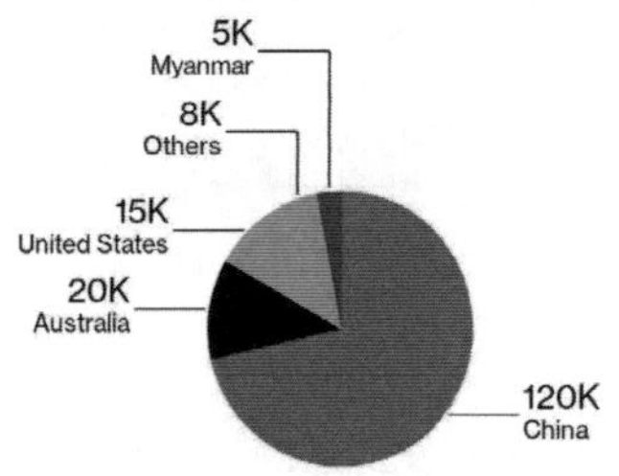

Source: U.S. Geological Survey, Mineral Commodity Summaries, February 2019
Data in metric tons of rare-earth oxide equivalent

BloombergOpinion

China Still Dominates Downstream RE Processing – Global Capacities

- Separation: 80%
- Reduction to Metals: 75%
- Magnet Manufacturing: 75%

JOC LLC

Rare Earths – Supply Demand (out of) Balance

Rare-Earth Elements
Originally produced for the October 2011 issue of *Scientific American*

What Are They Used For?

Scandium	Aerospace components, aluminum alloys
Yttrium	Lasers, TV and computer displays, microwave filters
Lanthanum	Oil refining, hybrid-car batteries, camera lenses
Cerium	Catalytic converters, oil refining, glass-lens production
Praseodymium	Aircraft engines, carbon arc lights
Neodymium	Computer hard drives, cell phones, high-power magnets
Promethium	Portable x-ray machines, nuclear batteries
Samarium	High-power magnets, ethanol, PCB cleansers
Europium	TV and computer displays, lasers, optical electronics
Gadolinium	Cancer therapy, MRI contrast agent
Terbium	Solid-state electronics, sonar systems
Dysprosium	Lasers, nuclear-reactor control rods, high-power magnets
Holmium	High-power magnets, lasers
Erbium	Fiber optics, nuclear-reactor control rods
Thulium	X-ray machines, superconductors
Ytterbium	Portable x-ray machines, lasers
Lutetium	Chemical processing, LED lightbulbs

Projected Supply & Demand (2016)

2016	Demand		Supply		Surplus
Element	tonnes	%	tonnes	%	(Deficit)
La	36,750	23.0%	52,000	26.7%	15,250
Ce	65,000	40.6%	81,000	41.5%	16,000
Pr	7,500	4.7%	9,500	4.9%	2,000
Nd	30,000	18.8%	31,500	16.2%	1,500
Sm	1,000	0.6%	3,750	1.9%	2,750
Eu	780	0.5%	500	0.3%	-280
Gd	2,225	1.4%	2,750	1.4%	525
Tb	450	0.3%	350	0.2%	-100
Dy	1,650	1.0%	1,450	0.7%	-200
Er	1,000	0.6%	800	0.4%	-200
Y	13,350	8.3%	10,000	5.1%	-3,350
Ho-Tm-Yb-Lu	250	0.2%	1,400	0.7%	1,150
TOTAL	159,955	100.0%	195,000	100.0%	35,045

Texas Mineral Resources: http://tmrcorp.com/ree/market_dynamics/

JOC LLC

Rare Earths – DOD Systems Need REE's

Not A Good Idea To Be Dependent On China

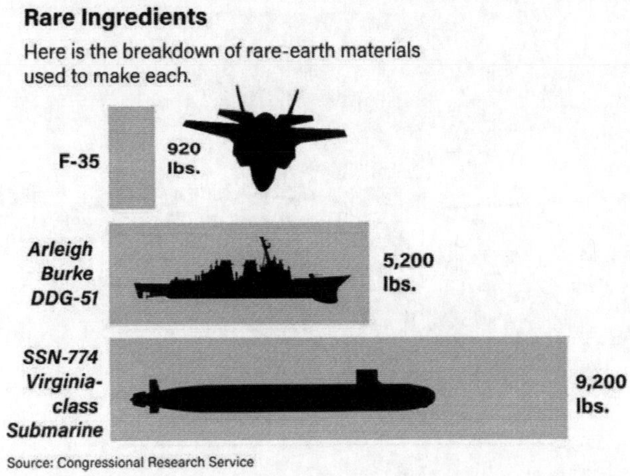

JOC LLC

Outline

- Introduction:
  - Me
  - Rare Earths
- The Ideal Permanent Magnet
  - Magnetics
  - Other Considerations
- Recent Developments
- Summary

JOC LLC

Hysteresis Measurement of Ideal Magnet

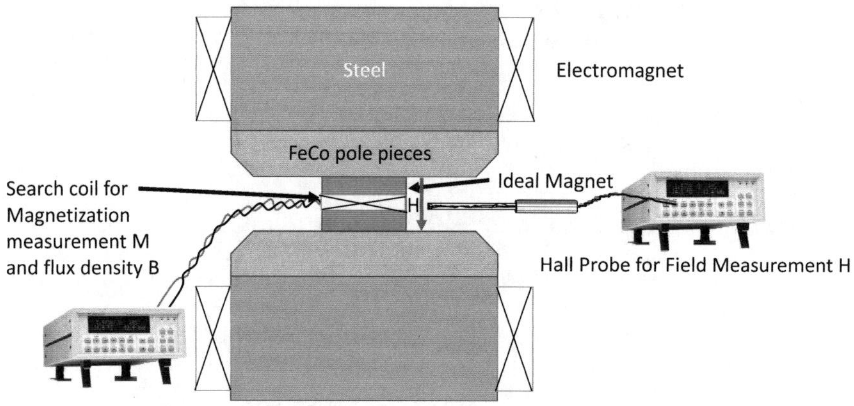

JOC LLC

Ideal Permanent Magnet -Magnetics

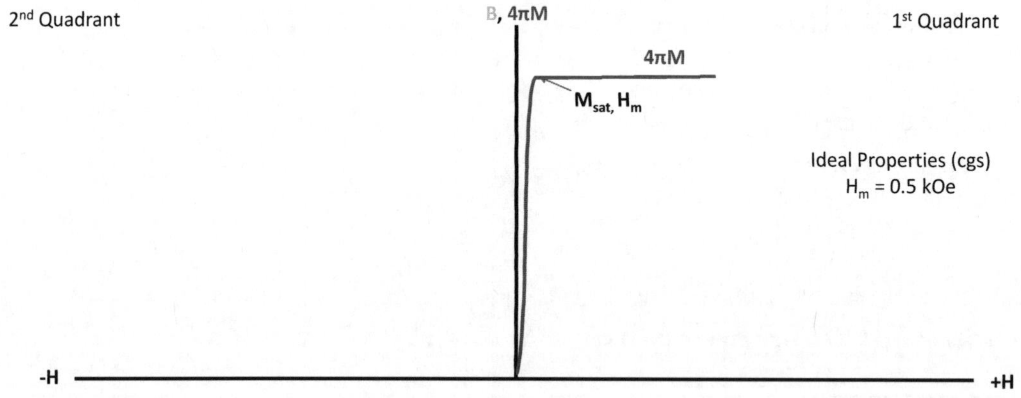

JOC LLC

Ideal Permanent Magnet -Magnetics

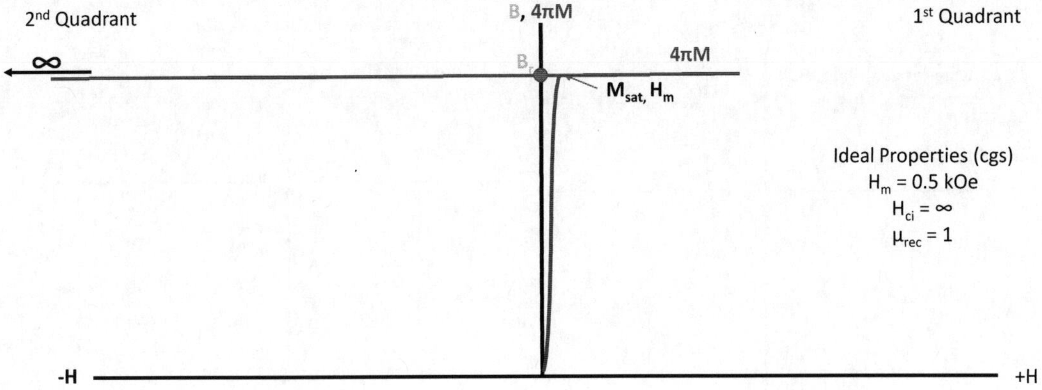

Ideal Properties (cgs)
H_m = 0.5 kOe
$H_{ci} = \infty$
$\mu_{rec} = 1$

JOC LLC

Magnetizing Field < H_c?
Grain Boundary Pinning Mechanism ($SmCo_5$ and $Nd_2Fe_{14}B$)

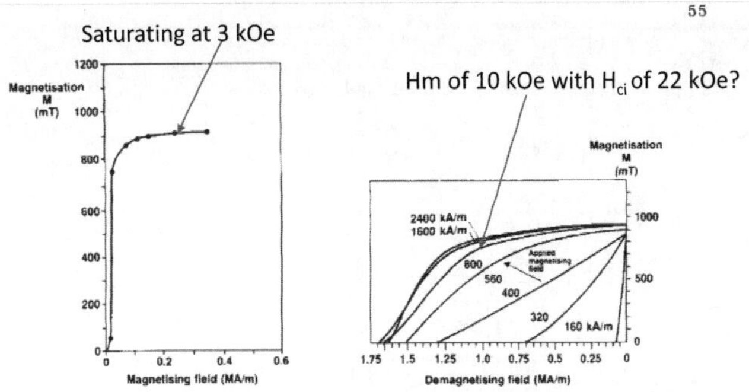

Fig. 1. Initial magnetisation curve of a thermally demagnetised $SmCo_5$ (RES 190) magnet.

Fig. 2. Dependence of coercivity on initial magnetising field of thermally demagnetised $SmCo_5$ (RES 190) magnets.

Source: J. Ormerod, Journal of the Less-Common Metals, 111 (1985) 49 - 69

JOC LLC

It's all About the Grain Boundaries (and Energy Wells)

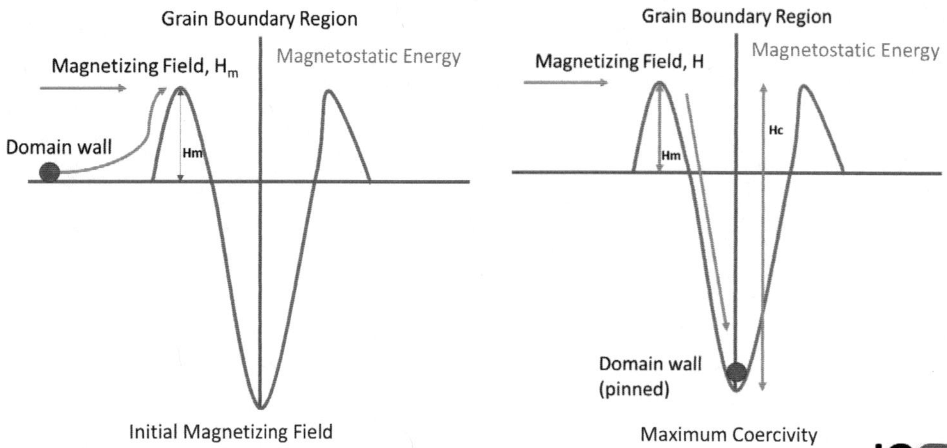

JOC LLC

Ideal Permanent Magnet -Magnetics

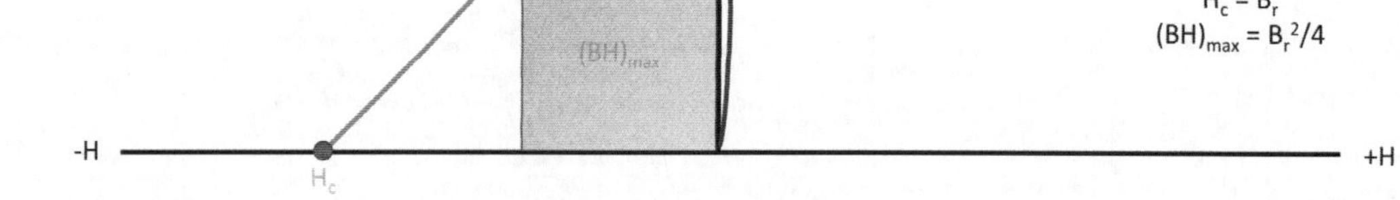

Ideal Properties (cgs)
H_m = 0.5 kOe
$H_{ci} = \infty$
$\mu_{rec} = 1$
$H_c = B_r$
$(BH)_{max} = B_r^2/4$

JOC LLC

Ideal Permanent Magnet -Magnetics

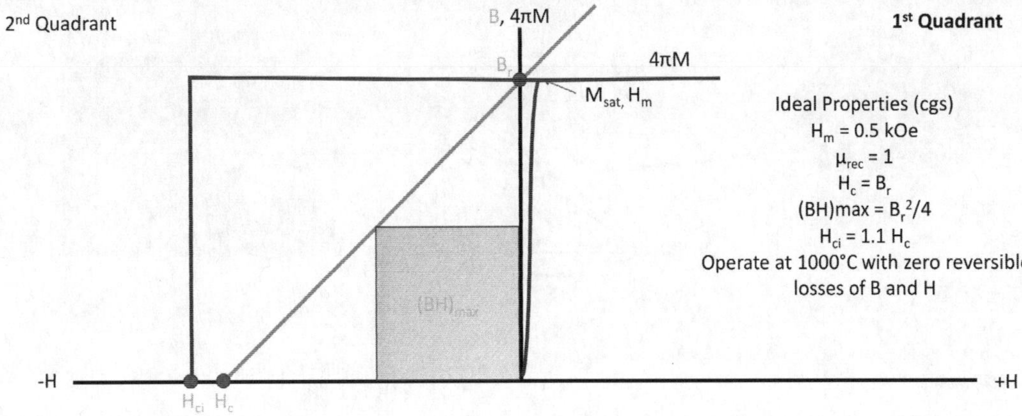

Ideal Properties (cgs)
H_m = 0.5 kOe
μ_{rec} = 1
$H_c = B_r$
$(BH)max = B_r^2/4$
$H_{ci} = 1.1\ H_c$
Operate at 1000°C with zero reversible losses of B and H

JOC LLC

Ideal Permanent Magnet -Magnetics

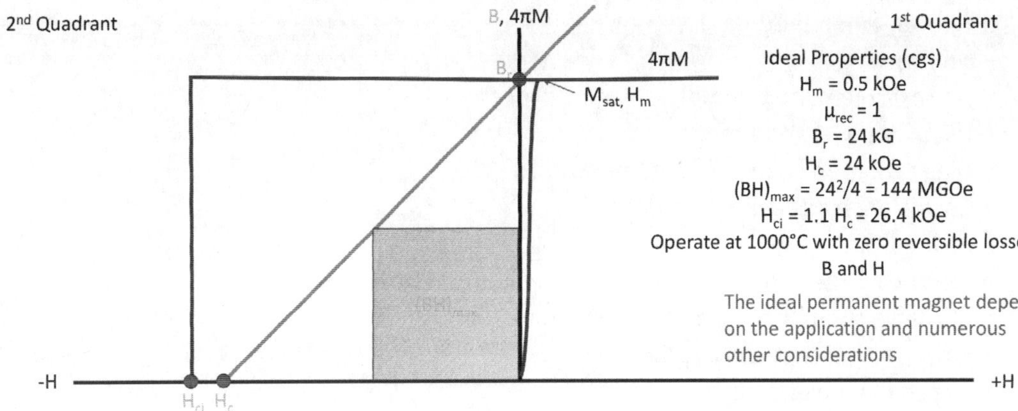

Ideal Properties (cgs)
H_m = 0.5 kOe
μ_{rec} = 1
B_r = 24 kG
H_c = 24 kOe
$(BH)_{max} = 24^2/4$ = 144 MGOe
$H_{ci} = 1.1\ H_c$ = 26.4 kOe
Operate at 1000°C with zero reversible losses of B and H

The ideal permanent magnet depend on the application and numerous other considerations

JOC LLC

The Ideal Magnet Depends On Application – Duh!

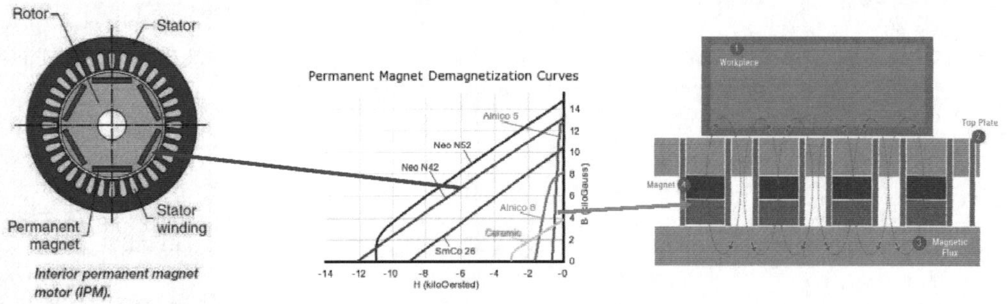

Sintered NdFeB IPM machines with a Br of 10 kG at 150°C with a straight-line B versus H characteristic.

Work holding application using alnico magnets with a B_r of 13 kG and H_c of 1.0 kOe

JOC LLC

The Ideal Permanent Magnet – Other Considerations

- Critical Material Content
- Uniformity
- No toxic or environmental damaging processes
- Usable temperature range
- Magnetization change with temperature (RTC)
- Corrosion resistance
- Physical strength
- Electrical resistivity
- Available sizes, shapes, and manufacturability
- Raw material cost and availability
- Net shape processing

JOC LLC

US DoE Critical Materials Assessment – 2015 - 2025

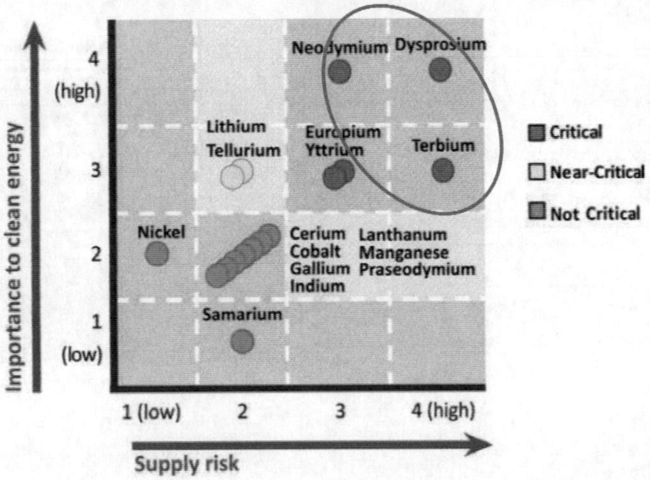

JOC LLC

EU 2020 Critical Raw Materials List

2020 critical raw materials (new as compared to 2017 in bold)		
Antimony	Hafnium	Phosphorus
Baryte	Heavy Rare Earth Elements	Scandium
Beryllium	Light Rare Earth Elements	Silicon metal
Bismuth	Indium	Tantalum
Borate	Magnesium	Tungsten
Cobalt	Natural graphite	Vanadium
Coking coal	Natural rubber	**Bauxite**
Fluorspar	Niobium	**Lithium**
Gallium	Platinum Group Metals	**Titanium**
Germanium	Phosphate rock	Strontium

JOC LLC

Dysprosium, Coercivity and Higher Application Temperatures

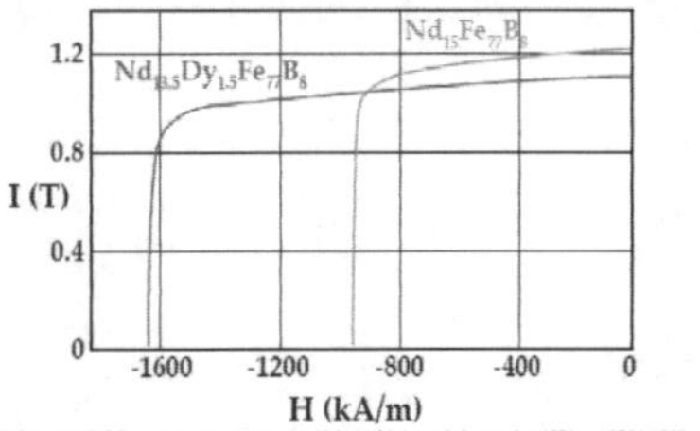

Reference: IEEE Transactions on Magnetics, Volume 20, Issue 5, September 1984, pp. 1584-1589

132 *Journal of Sustainable Metallurgy* (2018) 4:126–146

Table 2 Consumption of the different REEs in 2012, with the relative part of the total consumption being compared with the relative abundance of these elements in REE ores [33]

REE	Consumption in 2012 (tons of REOs)	Relative part of the total consumption (%)	Relative abundance (%)[a]
Lanthanum	31,495	27.8	24.9
Cerium	45,525	40.2	43.2
Praseodymium	4945	4.4	4.6
Neodymium	19,925	17.6	16.2
Samarium	515	0.5	2.2
Europium	425	0.4	0.3
Gadolinium	1020	0.9	1.4
Terbium	290	0.3	0.2
Dysprosium	845	0.7	0.9
Erbium	540	0.5	0.2
Holmium–thulium–ytterbium–lutetium	75	0.1	0.8
Yttrium	7650	6.8	4.9
Total	113,250	100	

[a]Average of 51 REE deposits

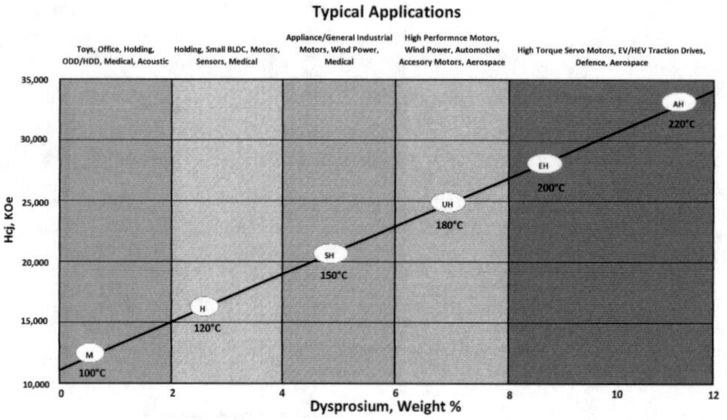

JOC LLC

Dy-Diffusion of NdFeB magnets

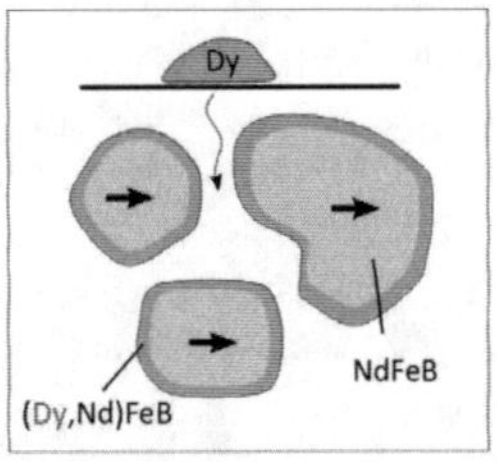

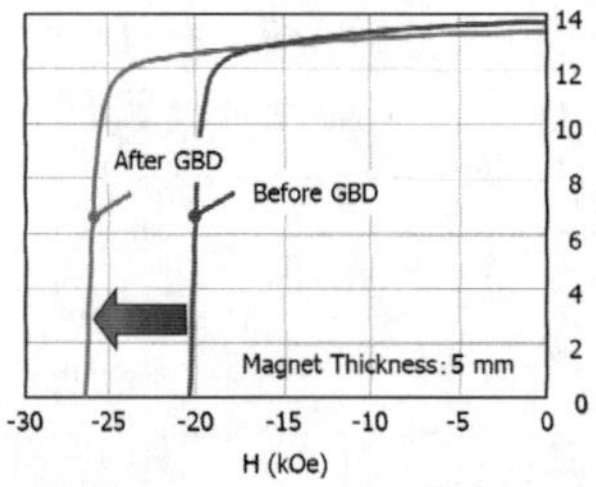

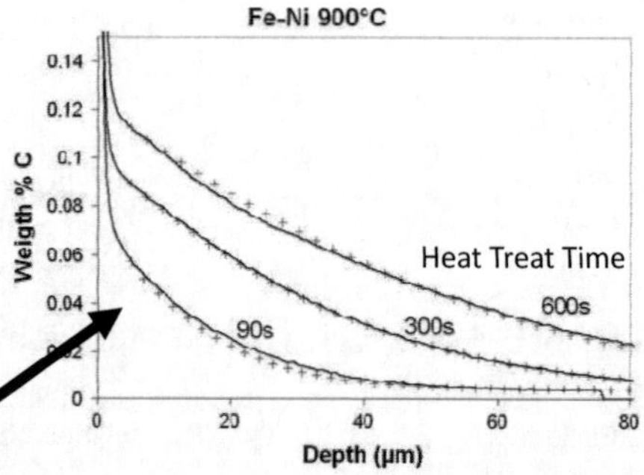

- Dy-diffusion technologies produce grades with approximately 50% Dy content and 5% improvement in Br.
- However, this is low temperature solid state diffusion process limiting the cross sectional that can be treated and tends to develop a concentration profile.

JOC LLC

Dy-diffused magnets in IPM EV drivetrains (GM Research Center)

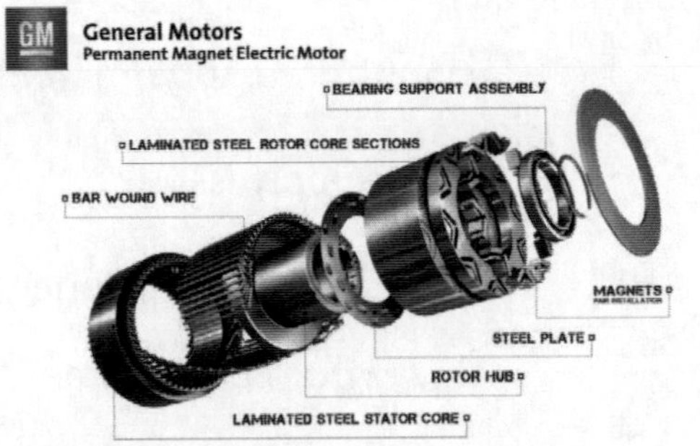

HEVs and EVs frequently have interior permanent magnet-type motors where the permanent magnets are embedded inside the rotor in one or more layers with V-shaped poles

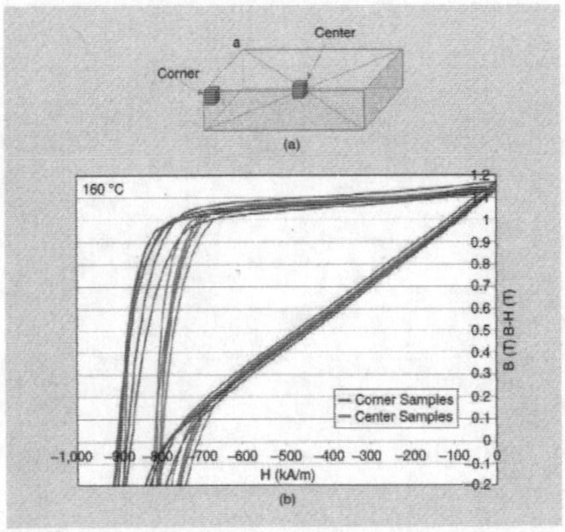

Figure 5. (a) A schematic of magnet locations of samples for VSM measurements. (b) The second quadrant BH curves for corner and center 2-mm cubic samples machined from magnets with Tb GBD measured by VSM.

M. P. Thompson et al, IEEE Electrification, March 2017

JOC LLC

Outline

- Introduction:
  - Me
  - Rare Earths
- The Ideal Permanent Magnet
  - Magnetics
  - Other Considerations
- Recent Developments
- Summary

JOC LLC

Achievement of High Coercivity in $Sm(Fe_{0.8}Co_{0.2})_{12}$

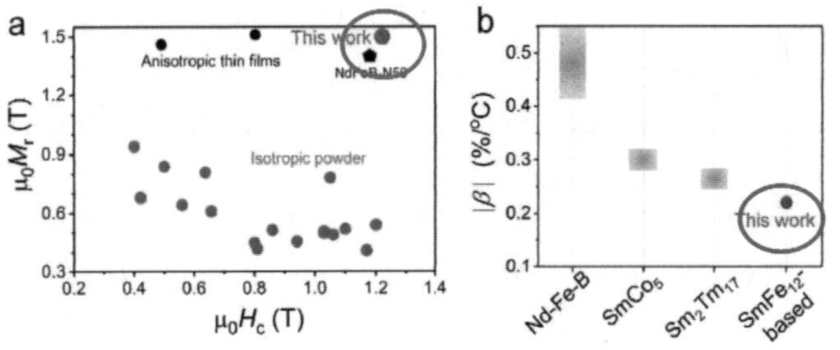

Acta Materialia, Volume 194, 1 August 2020, Pages 337-342

- Sputtered 100 nm films?
- Sm supply issue?
- Interesting but a long way from prime time.

JOC LLC

Outline

- Introduction:
  - Me
  - Rare Earths
- The Ideal Permanent Magnet
  - Magnetics
  - Other Considerations
- Recent Developments
- Summary

JOC LLC

New Permanent Magnet Materials
It's A Challenge To Hit The Sweet Spot

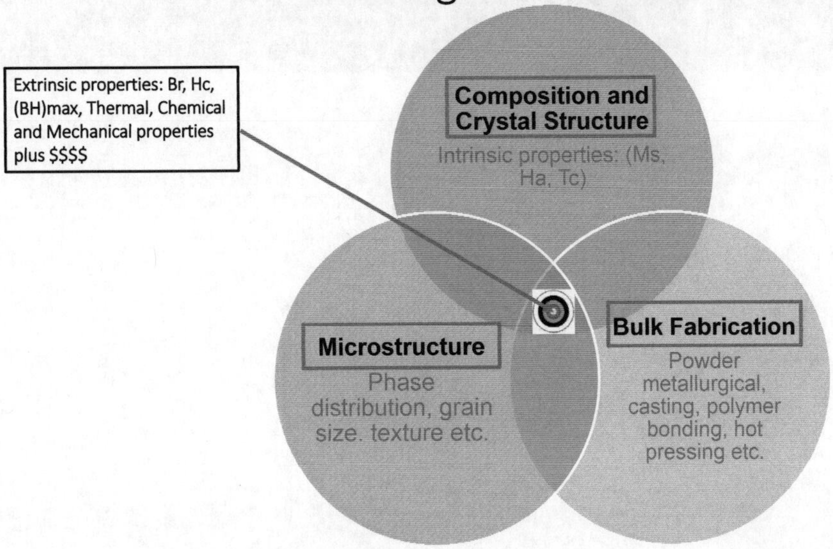

JOC LLC

Thank you for your attention
Any Questions?

Dr. John Ormerod

JOC LLC

Loudon, Tennessee

https://www.linkedin.com/in/jormerod/

+1 414 899 2859

Jormerod8@gmail.com

John@jocllc.com

JOC LLC

IOWA STATE UNIVERSITY
Department of Materials Science & Engineering

Prospects for Unconventional Sources of Magnet Materials

Alex King

Why do we need new sources?

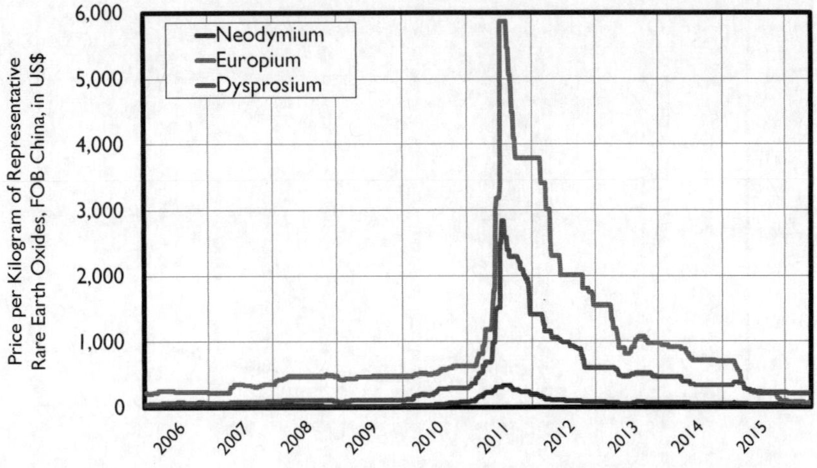

King: *Critical Materials*, Elsevier, 2020. Fig 1-6
Source of raw price data: Argus Media Inc. (direct.argusmedia.com)

IOWA STATE UNIVERSITY Department of Materials Science & Engineering 3

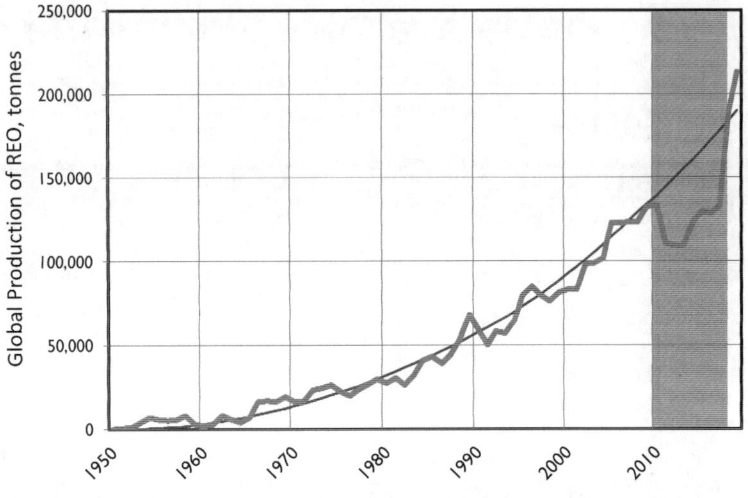

King: *Critical Materials*, Elsevier, 2020. Fig 4-1
Data from USGS Mineral Commodity Summaries

IOWA STATE UNIVERSITY Department of Materials Science & Engineering 4

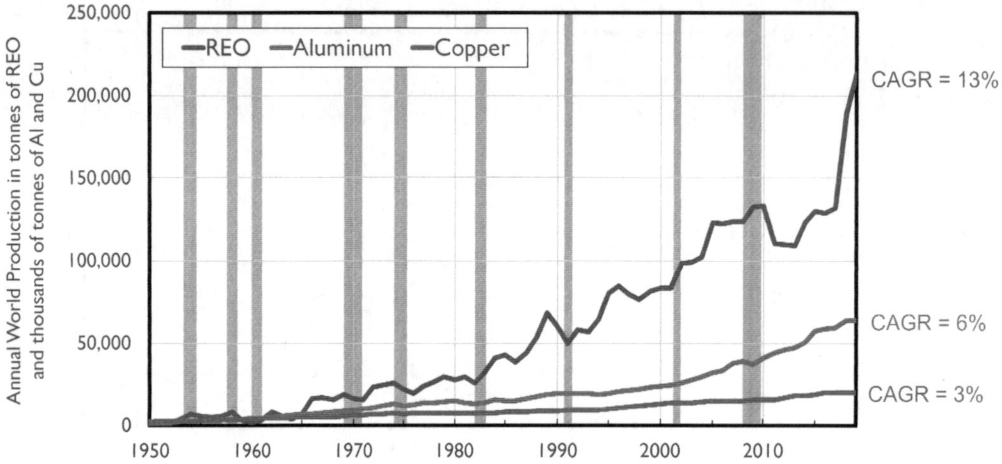

King: *Critical Materials*, Elsevier, 2020. Fig 1-7
Data from USGS Mineral Commodity Summaries

What sources may be available to meet the need?

- Conventional sources
  - Geologic mines

What sources may be available to meet the need?

- Conventional sources
  - Geologic mines
- Unconventional sources
  - Recycling
  - Co-production
    - Coal mines
    - Phosphate mines
  - Aqueous sources
    - Dissolved materials
    - Seabed deposits
  - Space

IOWA STATE UNIVERSITY Department of Materials Science & Engineering 7

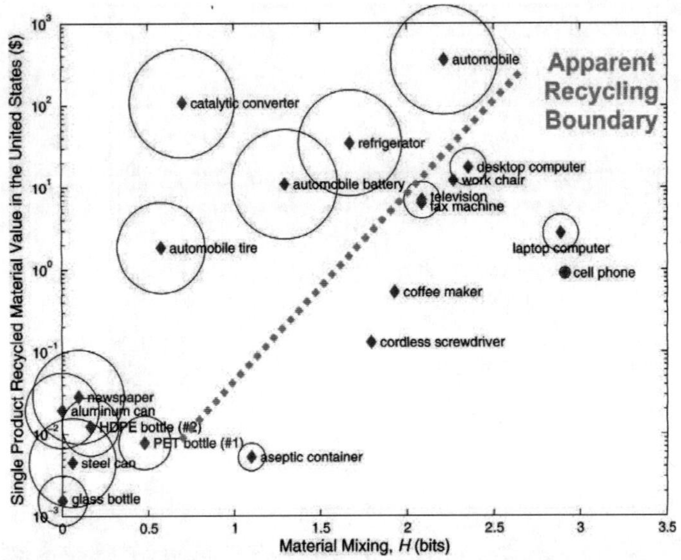

King: *Critical Materials*, Elsevier, 2020. Fig 7-1

Material value and device complexity set limits on what gets recycled.

Original figure from J.B. Dahmus, T.G. Gutowski, What Gets Recycled: An Information Theory Based Model for Product Recycling, *Environmental Science & Technology*, 41 (2007) 7543-7550. © 2007, American Chemical Society. Used with permission.

IOWA STATE UNIVERSITY Department of Materials Science & Engineering 8

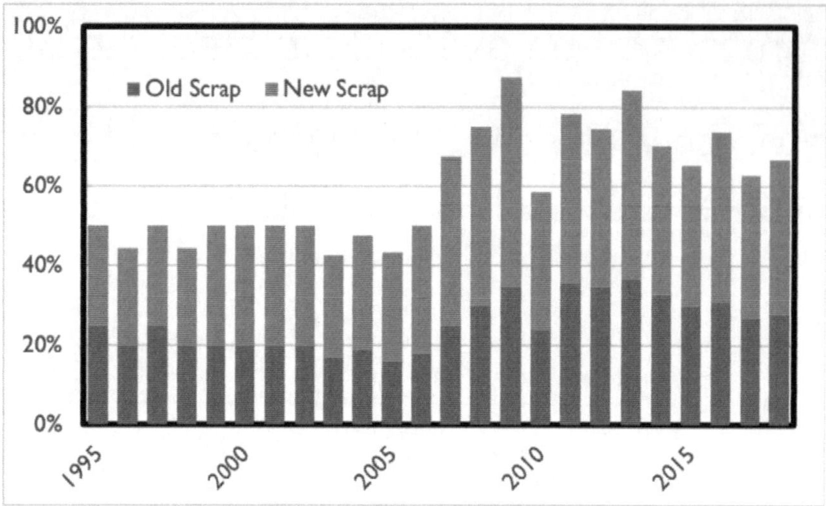

King: *Critical Materials*, Elsevier, 2020. Fig 7-2
The fraction of total US annual aluminum demand that has been met by recycled material, from 1995 to 2018.

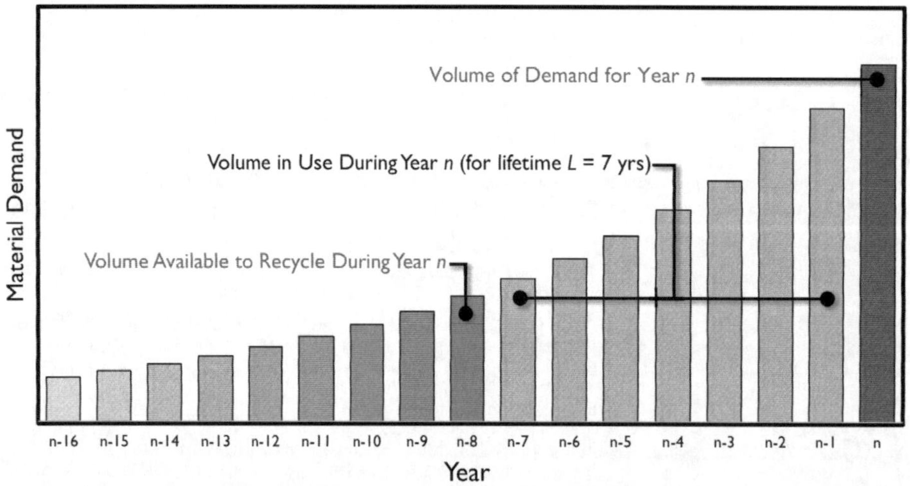

King: *Critical Materials*, Elsevier, 2020. Fig 7-3

How much of the need can be met by recycling?

$$F_R = E_c E_e / (1+G)^{L+1}$$

- F_R = fraction of current need that can be met by recycling
  - E_c = efficiency of collection
  - E_e = efficiency of extraction
  - G = annual rate of demand growth
  - L = product lifetime, in years

How much can recycling contribute?

Material	E_c	E_e	G	L	F_R
Aluminum	50%	100%	6%	1	44%
REE	30%	99%	13%	6	13%

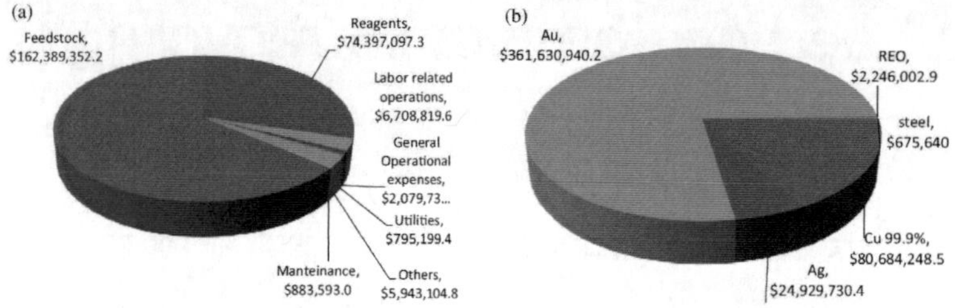

Fig. 3. Operational (a), and revenue (b), cost distribution for the second year of operation for the BCS process.

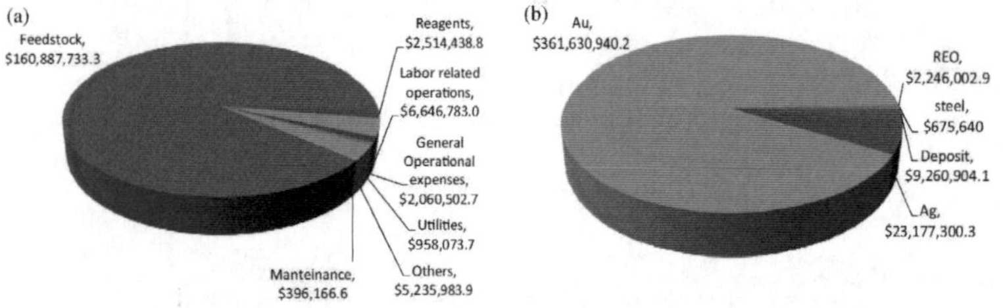

Fig. 4. (a) Operational and (b) revenue cost distribution for the second year of operation of the ER process.

- Comparing costs and revenues for two different processes for recovering metals from electronic scrap.
- Note, *inter alia*, that feedstock acquisition is, by far, the biggest cost for either process.

L.A. Diaz and T.E. Lister, *Waste Management* **74** (2018) 384-392

Recycling: summarizing comments

- The rapid growth of demand for REEs limits the contribution that recycling can make.
- For old scrap, recovery is the biggest single cost. Focus on products that are easier to recover.
- Where possible, focus on less complex products.
- Focus on products with smaller demand growth.
- Focus on products with shorter lifetimes.

Co-production

- REEs pass through phosphate fertilizer production processes at a rate equivalent to current global demand.
- Coal mine gangue can contain REEs.
- In both cases the concentrations are in the range of a few hundred ppm. *They are dilute sources.*

IOWA STATE UNIVERSITY Department of Materials Science & Engineering 15

The logistical challenge of dilute sources

Source Concentration	500 ppm	1,000 ppm	100,000 ppm
Tonnes of feedstock required to produce 1 t of TREO	2,000	1,000	10
Truckloads (of 20 t each) required to produce 1 t of TREO	100	50	0.5

IOWA STATE UNIVERSITY Department of Materials Science & Engineering 16

The logistical challenge of dilute sources

Source Concentration	500 ppm	1,000 ppm	100,000 ppm
Tonnes of feedstock required to produce 1 t of TREO	2,000	1,000	10
Truckloads (of 20 t each) required to produce 1 t of TREO	100	50	0.5

The production of a modest 10,000 t of REO per year would require one truckload of feedstock, *plus* all of the chemicals needed to process it, *plus* the removal of all of the waste, roughly *every minute of every working day*.

IOWA STATE UNIVERSITY   Department of Materials Science & Engineering 17

More detail about these and other unconventional sources:

Critical Materials, 1st Edition

Alexander King

Elsevier, 2020

Paperback ISBN: 9780128187890
eBook ISBN: 9780128187906

https://www.elsevier.com/books/critical-materials/king/978-0-12-818789-0

IOWA STATE UNIVERSITY   Department of Materials Science & Engineering 18

Thank you

Questions?

IOWA STATE UNIVERSITY Department of Materials Science & Engineering 19

Roskill

RE supply impacting the development of RE permanent magnet demand in key applications

Magnetics 2021
January 19-20

Leslie Liang
Senior consultant, Roskill

Disclaimer

The statements in this presentation represent the considered views of Roskill Information Services Ltd. It includes certain statements that may be deemed "forward-looking statements". All statements in this presentation, other than statements of historical facts, that address future market developments, government actions and events, are forward-looking statements. Although Roskill Information Services Ltd. believes the outcomes expressed in such forward-looking statements are based on reasonable assumptions, such statements are not guarantees of future performance and actual results or developments may differ materially from those in forward-looking statements. Factors that could cause actual results to differ materially from those in forward-looking statements include changes in general economic, market or business conditions.

While Roskill Information Services Ltd. has made every reasonable effort to ensure the veracity of the information presented it cannot expressly guarantee the accuracy and reliability of the estimates, forecasts and conclusions contained herein. Accordingly, the statements in the presentation should be used for general guidance only.

Roskill

Outline

- Rare earth operations in China
- China seeking Rare earth feedstock globally
- ROW speed up RE layout
- High strength magnet applications in NdFeB
- RE market/elements balance

Roskill

China rare earths supply

- China RE industry is consolidated in 2014 formed in 6 SOEs
- Each SOE dominate one or multiple province's RE resources
- China Light REEs production is sourced from in Inner Mongolia, Sichuan and Shandong - Bastnaesite deposits
- China Heavy REEs production is sourced from Guangdong, Guangxi, Fujian, Hunan, Yunnan- Ion adsorb clay (IAC) deposits

China Rare earth layout by 6 SOEs

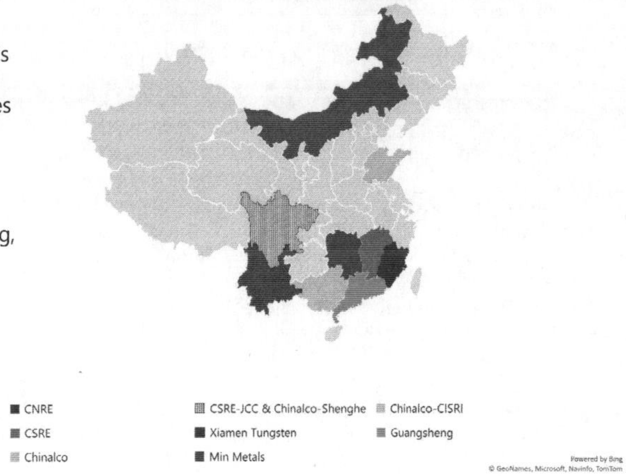

Roskill

Source: Roskill

China towards environmental friendly

- Since 2010, China has continuously strengthened its environmental legislation
- A number of state and regional inspections is implemented to reduce air, earth, solid waste and water pollution
- Environment legislation forced IAC deposits halted since 2016 (all HREE elements effected)
- Prospecting and mining on single monazite deposit was banned in 2011 (effect Pr,Nd supply mostly)

Chinese environmental legislation

- <Guiding Catalogue for the Elimination of Outdated Production Process Equipment and Products in Certain Industrial Industries (2010)>
- <Pollutant Emission Standard for Rare Earth Industry> (2011)

Introduction of Chinese environmental legislation and amendments since 2015

January 2015	Environment law introduced
January 2016	Atmospheric pollution prevention law amended
September 2016	Law of EIA (Environment impact assessment) introduced
January 2017	Water pollution prevention law amended
January 2017	Interpretation by the Supreme People's Court & Procuratorate on application of law in handling EV pollution criminal cases released
November 2017	Introduction of interim environmental protection policy for construction projects
January 2018	Trail establishment of administration permitting pollution discharge
August 2018	Public participation in EIA begins
December 2018	Market access negative list
April 2019	Guidance catalogue for industrial structure adjustment (2019 version draft)
Dec 2019	Notice on the continued strengthening of the order of the rare earth industry
June 2020	Foreign investment access negative list (2020 version)
Dec 2020	Export control Act

Source: Roskill

Roskill

China Rare earths supply targeted LREEs

- LREE Quota increased 39% in 3 years from 2017 to 2020
- Resources tax rate on LREEs various from 7.5% to 11.5%
- LREE refined plants running at high operational rates
- CNRE Baiyun Obo mine: the largest LREE producer （Inner Mongolia）
- **Sichuan** accounted for the dramatic increasing
  - CSRE/Jiangxi Copper (JCC) - Maoniuping
  - Chinalco/Shenghe – Dalucao
  - Other potential mines in the region
- Chinalco/CISRI – Weishan: a minor increase （Shandong）

LREE Mine quota uplift
LREE Refined output ramp up
Authorities encourage LREEs deposits operational

China Rare earth Mine quota (t REO)

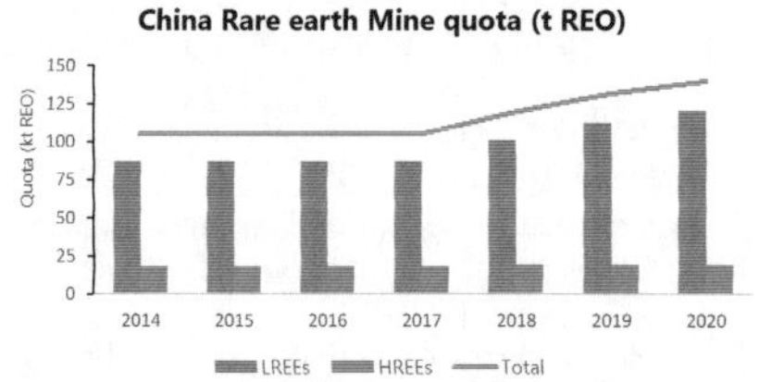

China LREEs quota by SOEs/deposits 2017 (Inside) Vs 2020 (Outside)

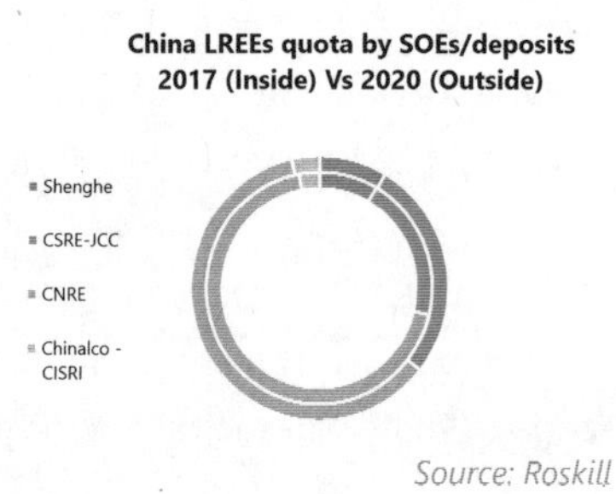

Roskill

Source: Roskill

China IAC disruptions Vs Potential developments

- Environmental legislation is the key reason to IAC disruptions
- Resources tax rate on HREEs reduce to 20% from 27% in Sept. 2020
- HREE refined plants running at low operational rates

- IAC mine prospects?
  - CSRE re-assumption (Jiangxi)
  - Min-metals Jianghua (Hunan)
  - Guangsheng Huaqi (Guangdong)
  - Guangsheng Xinfeng（Guangdong）
  - Chinalco Huashan (Guangxi)

HREE Mine quota flat
HREE Refined output disruptions
Authorities encourage HREEs import

China Quota by 6 SOEs, 2020 （t REO)

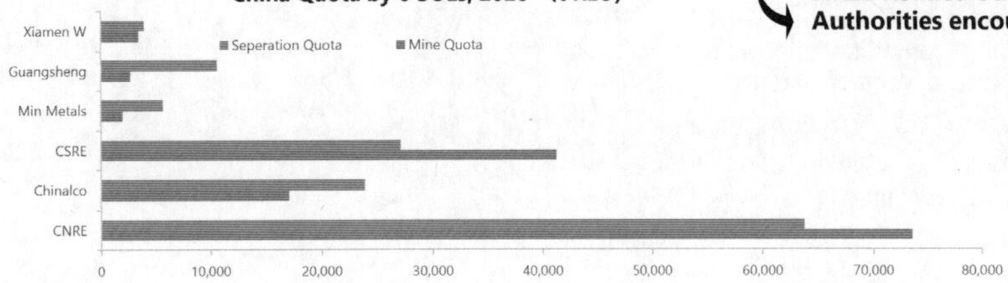

Roskill

Source: Roskill

Rare earth industry is undergoing a significant change

- Chinese mine supply averaging 97% of global production between 2005 and 2010
- **China** accounted for approximately 57% of global production in 2020, down from 82% in 2017
- Reduction in **illegal REE sources** has been significant as a result of price pressure, **government crackdowns** & **consolidation**
- Environmental legislation and disruptions to domestic supply chains incentivise diversification of feedstock sources
- China continue to dominating rare earths supply in 2020
- ...but diminishing importance

World: Mine production of Rare earths, 2005-2020 (t REO)

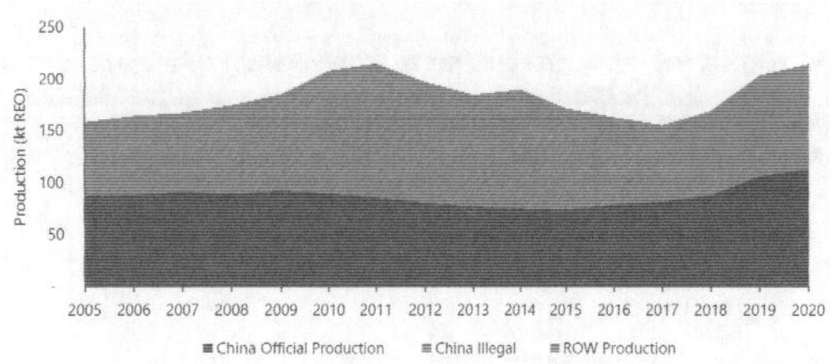

China: Contribution of illegal mine production of rare earths, 2005-2020 (t REO)

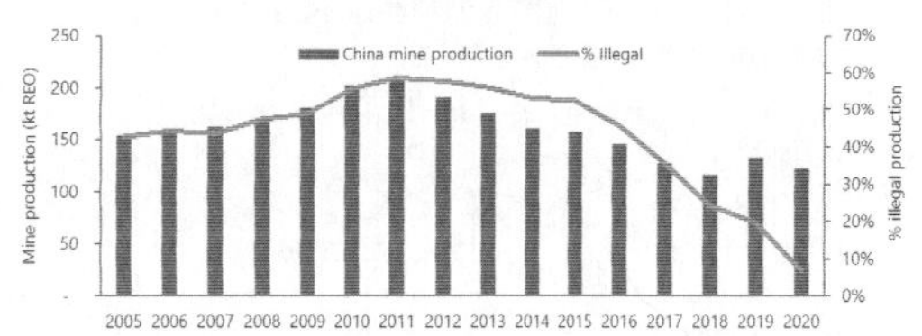

Roskill

Source: Roskill

China is seeking resources from ROW

HREEs

- HREE supply rely on 3rd parties (Imported)
- Limited (viable & competitive) alternatives for HREE supply ex-China
- Myanmar feed import bans in 2019 H1 and resumed in H2
- Import from Myanmar stays high in 2020
- But in forms of different code-MREC and unlisted

Monazite

- Domestic monazite producers turn to treat import feed
- In the form of Zr, Ti and Th sand (content various)
- Mainly from Thailand, Vietnam, Madagascar, Australian and Brazil
- Government encourages monazite import

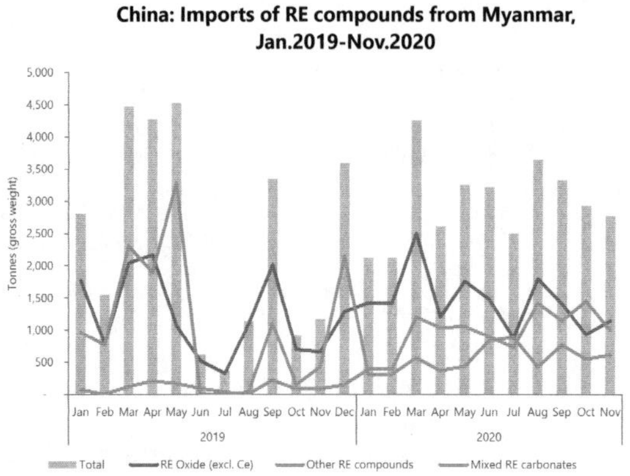

Roskill

Source: Roskill

ROW: More weighted in rare earth mining

- Global RE mine production reached 154.5kt REO in 2017, reach 214kt REO in 2020
- Non-Chinese RE mine production accounting for 43% in 2020
- MP (USA) restart of mining operations- second largest rare earth producer after CNRE(China Northern Rare Earths).
- Lynas (Australia) continued ramp-up and expansions at Mt. Weld
- Other new and existing production in India, Russia, Burundi, Brazil and by-product rare earth mining in Australia and Madagascar
- HREE mine production heavily concentrated in China

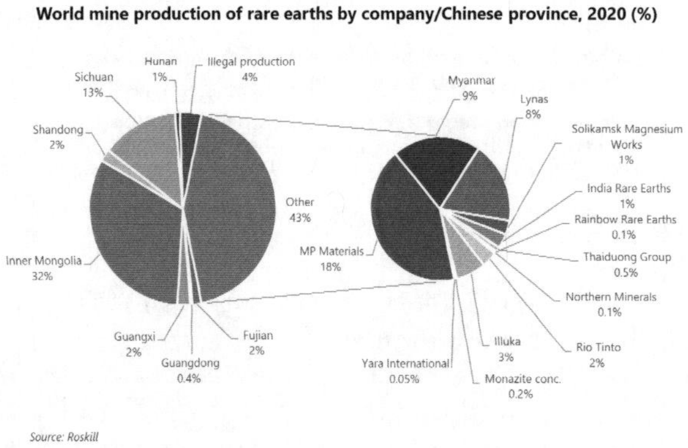

Roskill

Source: Roskill

ROW: Constrained by Refined production

- Refined production of RE is heavily centred within China
- With domestic production being supported by both domestic mining and imports of mineral concentrates and ores materials
- China accounting for 85% of global production in 2020
- Tech/experience personal/R&D dominate by China
- Prospecting growing refined supply from the USA, EU and Australia

World refined production of rare earths by company/Chinese province, 2020 (%)

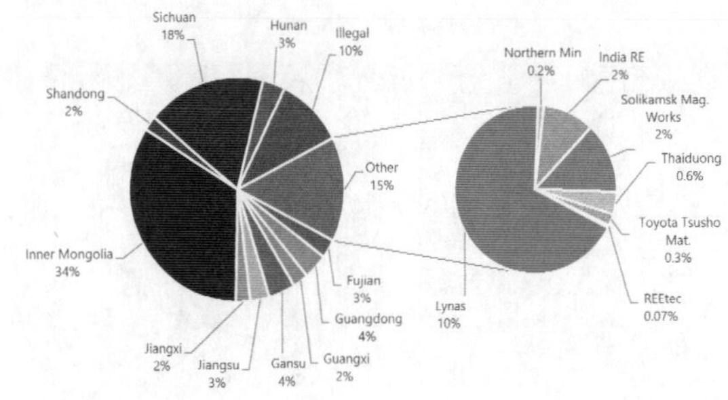

Source: Roskill

Roskill

Source: Roskill

China SRB storage Dilemma?

- State Reserve Bureau (SRB) primary function is to stabilize prices
- Fanya released stock in 2019, all off taken by Min-metals
- From 2016 to 2016, SRB took 4 stockpile, totalling 5.8kt REO.
- SRB's intentions to increase stockpiles, plan was made in 2016
- Six SOEs have been separately made stockpiles on different Rare earth products, waiting for SRB to restocks. （Specially on Pr,Nd,Dy,Tb,）

SRB storage plan 2016 (t REO)

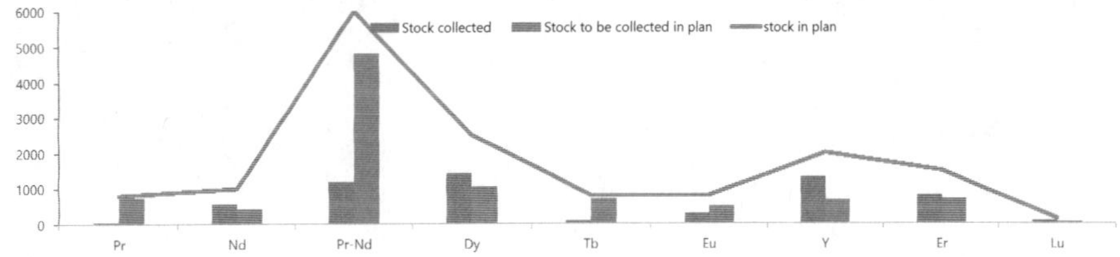

Roskill

Source: Roskill, Baiinfo

What are rare earths? A quick overview of a diverse market

- Suite of **15 elements** known as the lanthanides including yttrium that naturally occur together.
- Used across **high-tech and new energy applications** to improve performance and efficiency.
- Electric vehicles are focussing the attention on **neodymium** (Nd) and **praseodymium** (Pr).

Which rare earths are used where?

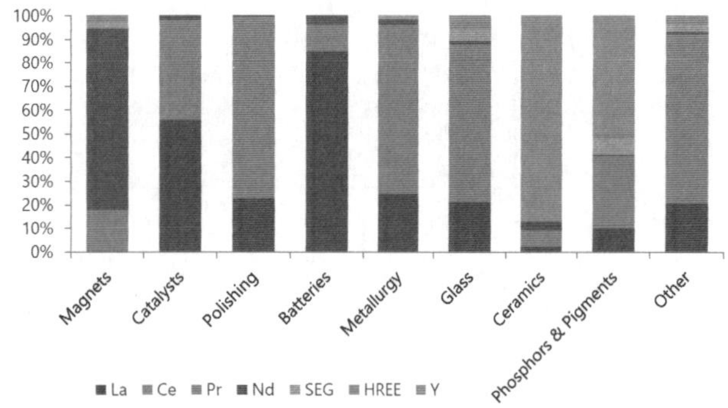

Rare earth consumption by application 2020

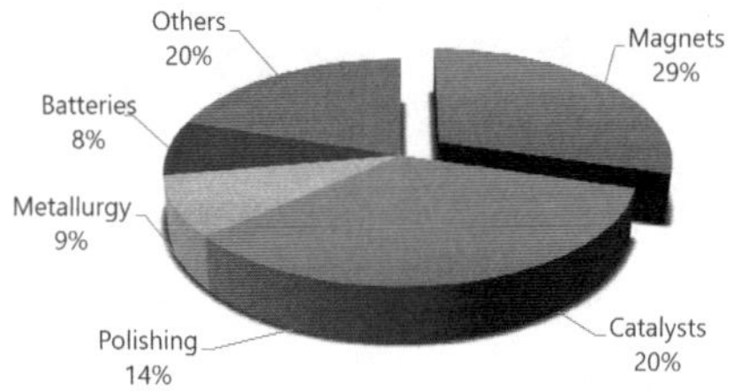

Roskill

Source: Roskill

Drivetrain motors require higher quality and larger PM

- **Permanent magnet** synchronous motors (PMSM) are the motor of choice in electric vehicle drivetrains (> 90%).
- **Drivetrain motors** require higher quality and larger permanent magnets compared to traditional applications
- Pure battery EVs and LSEVs use more permanent magnets per motor than HEVs

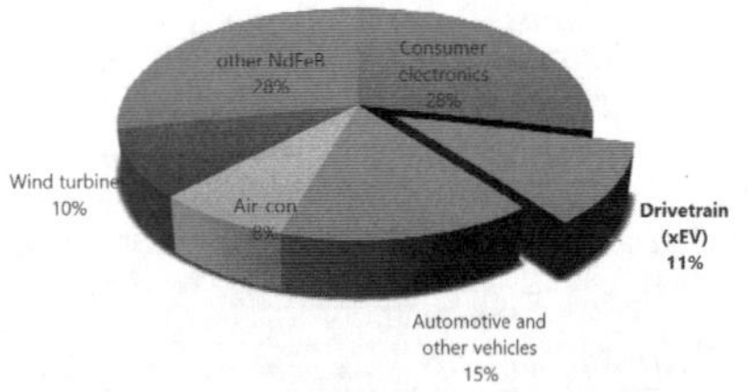

NdFeB Applications 2020

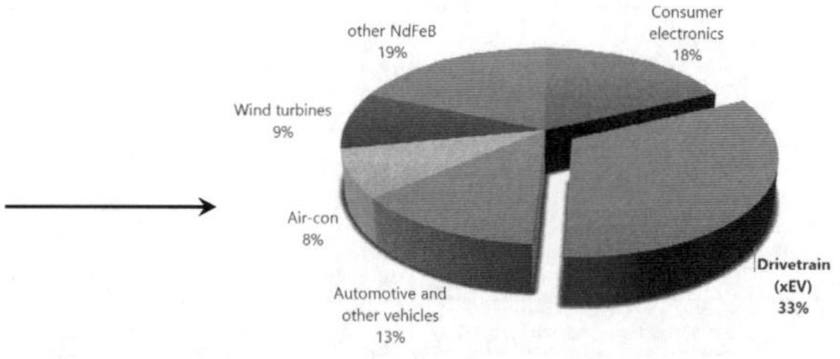

NdFeB Applications 2030

Roskill

Source: Roskill

NdFeB: technology of choice

- Leading power-density permanent magnets are rare earth **neodymium-iron-boron** (NdFeB) - highest power-to-weight ratio
- REPMs allow for the highest energy efficiency rating in magnet motors/generators
- Resulting benefits: Performance, Miniaturisation, Cost saving ,Maximising gains from high-cost investment
- RE Permanent magnet: Electric vehicles are **focus on Nd, Pr and Dy (Tb)**

Comparison of motor efficiency

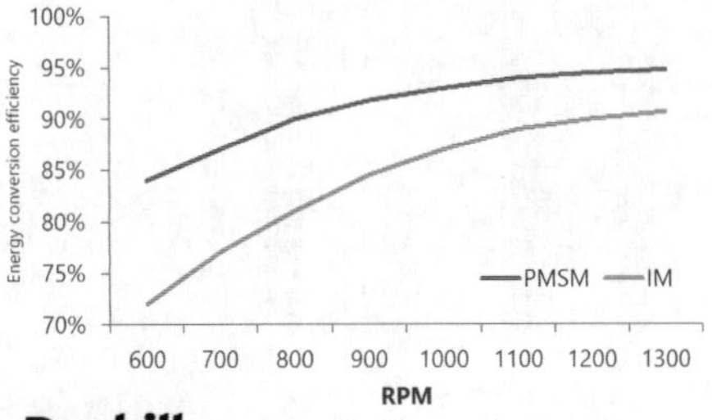

Choice of motor technology in NEVs

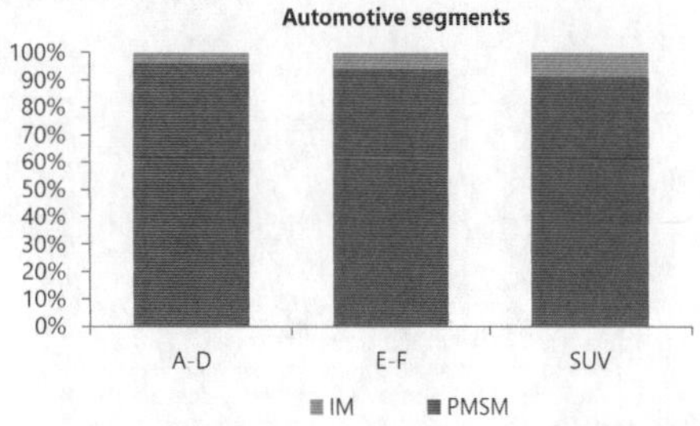

Roskill

Source: Roskill

NdFeB in key applications

- xEV: China 2020-2035 NEV plan
- AC converter (energy consumption)
  - China legislation: Air-con energy efficiency uplift require (2020 Jun.)
  - Current: AC fixed Vs AC invertor
  - Prospect: 45% of current AC (~4ML units AC) to be eliminated from Y2020
- Wind turbine(energy generation)
  - Chinese subsidy withdrawal from Jan. 2021
  - Current: Double fed motor take the main share >70%, PMSM 30%;
  - Prospect: On-shore decline Vs off-shore strong growth installation in between 2025 and 2030
- Consumer electronics
  - Miniaturisation – phones
  - JLMag : mass production in 3C field

China AC sales by energy efficiency grade (%)

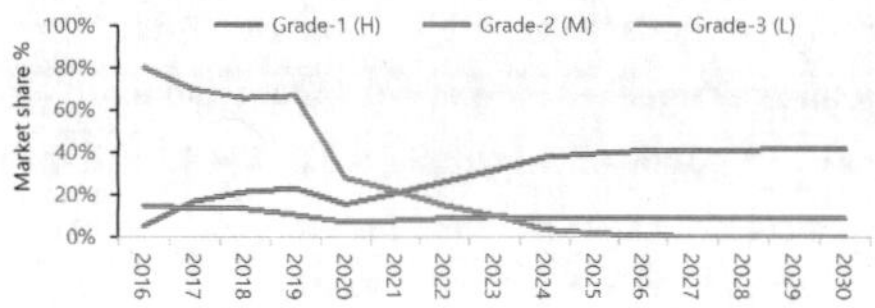

Global NdFeB demand in key sectors (t)

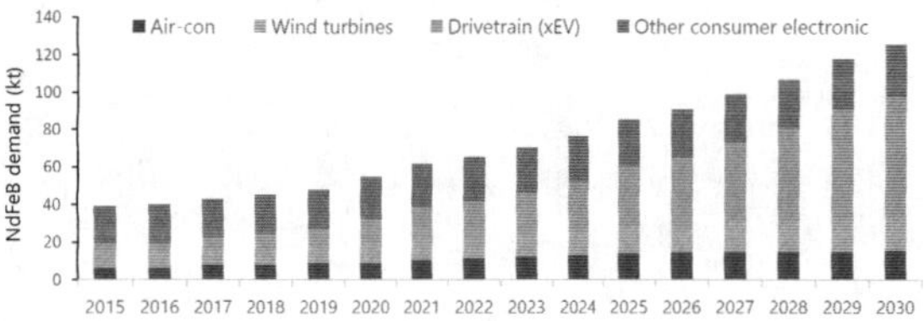

Roskill

Source: Roskill, ChinaIOL.com

Rare earth recycling

- Over 50 operations and projects for rare earth recycling in China
- Recycling in China is centralized in 2020
    - Mainly located in Jiangxi and Shandong
    - Capacity of close to 300ktpy
    - Recycling focussed on NdFeB
    - Mainly destructive dis-assembly
- ROW recycling remains limited
- Production of RE from secondary sources is expected to growth strongly

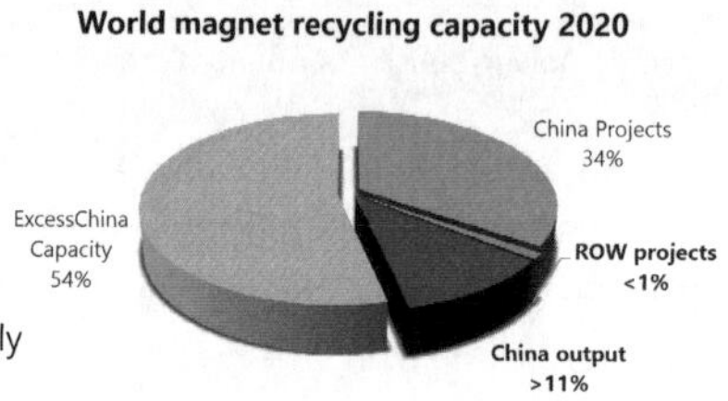

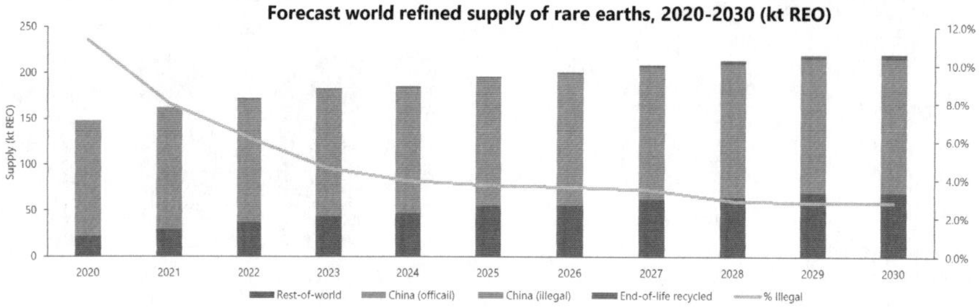

Roskill

Source: Roskill

Market Balance outlook

- The overall market balance remain in a significant and growing oversupply
- The increase is largely caused by the overproduction of lanthanum (La) and cerium (Ce)
- Producers target less naturally abundant rare earth elements (Nd,Pr,Dy - RE magnetic elements).
- **For every tonne of Nd supply, we get: 4.4 t other REE (Mainly Ce, La)**
- The increase in secondary production of rare earths is expected to reduce the TREO: Nd production ratio

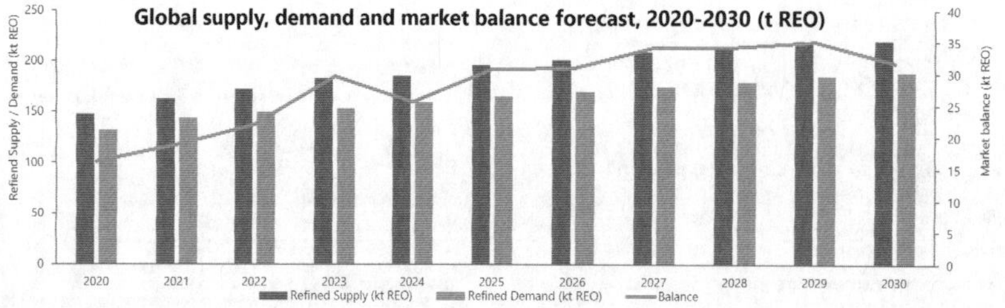

Roskill

Source: Roskill

Summary: Rare earth supply/demand balance

- Rare earth market defined by neodymium supply
- Price trend divergence between magnet and non-magnet materials - continued oversupply of LaCe
- Dy (& Tb) supply tightness more sensitive than other magnet materials

Forecast proportional consumption of rare earths, 2020-2030 (supply/demand)

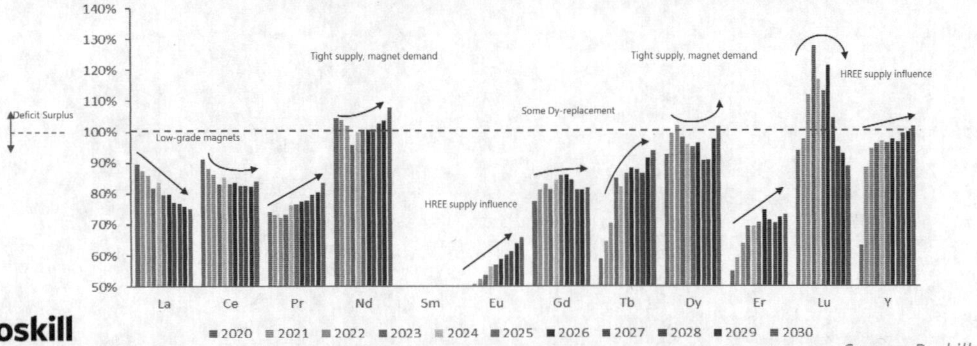

Roskill

Source: Roskill

20

MARKET REPORTS – COSTS & DATA – CONSULTING – EVENTS

Roskill

Leslie@roskill.com
Senior Consultant

matesy
Magnetic Technologies & Systems

Matesy GmbH
Magnetic Technologies and Systems

Inline inspection systems for the quality control of permanent magnets and magnet assemblies

Benjamin Wenzel, Magnetics Conference, January 20th 2021

Content

- **About Matesy**
- **CMOS magview:** Magneto-optical systems for industrial applications
- **M-axis:** Inline measurement of magnetic properties
- **MHLS:** Matesy Hall Line Sensor for inline inspections of permanent magnet assemblies
- **Conclusion**

2

The Matesy GmbH
Quality made in Jena

- Is an innovative technology company located in the center of the university town of Jena
- The company was founded in 2008 as a spin-off of the R&D institution "INNOVENT Technology Development"
- Specialized in the visualization, characterization and generation of magnetic fields and twist testing
- We develop and produce systems and technologies of the highest standard
- We are service providers (order measurements)
- We work together with the National Metrology Institute of Germany - the Physikalisch-Technische Bundesanstalt for calibration, reference magnets and the development of new standard measurement methods

3

The Matesy GmbH - Solutions around magnetism, magneto-optics and optics

Our core competencies are

- The visualization and measurement of magnetic fields
- Magnetic measuring and calibration systems
- 3D tracking of magnetic markers
- Customized Helmholtz Coils (latest 1.4 m diameter 3D)
- Shaft twist testing

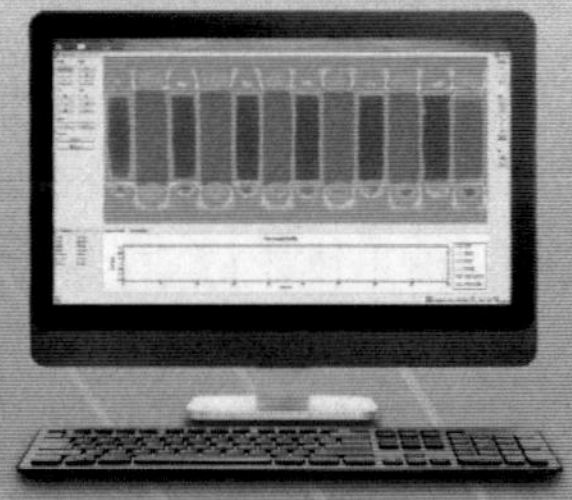

4

Industries

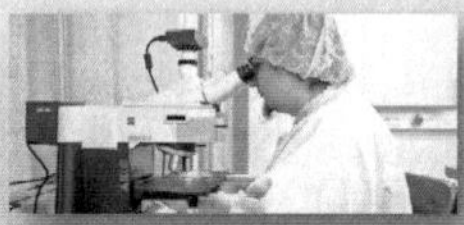

Automotive & Engineering

Our devices & systems offer you the possibility:

- to test magnets and magnetic components accurately
- to test materials for their magnetic properties
- to detect twist structures and roughness on shafts in seconds

Glass & Solar Industry

For the glass and solar industry we offer systems for:

- accurate measurement of magnet bars and magnetrons
- measurement of targets or pipe targets
- surface inspection of flat glass coatings

Forensics

For forensic investigations, we ran:

- different types of magnetic field sensors
- the cmos-magview family (magnetic field camera systems)
- the mageye and mageye for mobile surface analysis
- adapter with corresponding sensors for microscopic surface analysis

Science & Research

For scientific applications we build devices:

- for generating magnetic fields
- to visualize magnetic fields and magnetic structures
- for magnetic monitoring

5

The Matesy GmbH - Magnetic measuring and calibration systems

- Determination of permanent magnetic properties (magnetic moment, remanence, etc.)
- Generation of magnetic fields
- Determination of magnetic information and geometries of magnet bars and tubular targets

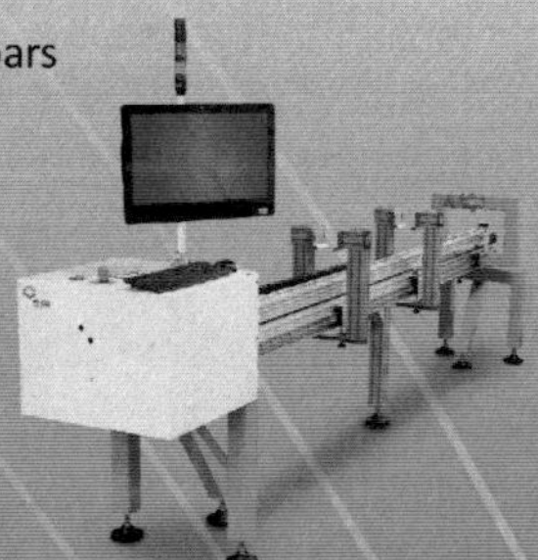

6

The Matesy GmbH - Magnetic field visualization

Our systems produce:

- Magnetic field images of the near surface stray fields

Application:

- Forensic science
- Material characterization
- Quality control

7

The Matesy GmbH - Magnetic field visualization / magneto-optics

Matesy GmbH offers equipment for :

- Visualization of magnetic structures and fields (quality inspection, crack detection)
- Checking security features of e.g. banknotes
- Recognition of manipulated serial number (for example in weapons)
- Representation of domain structures in electrical sheets

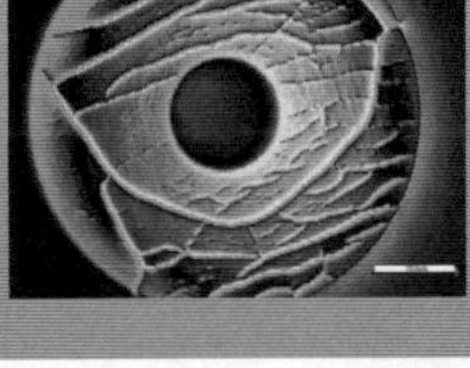

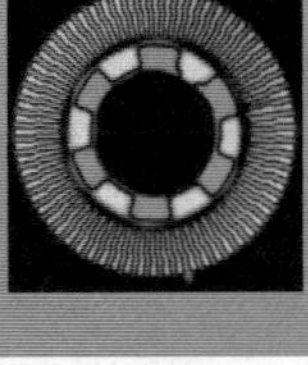

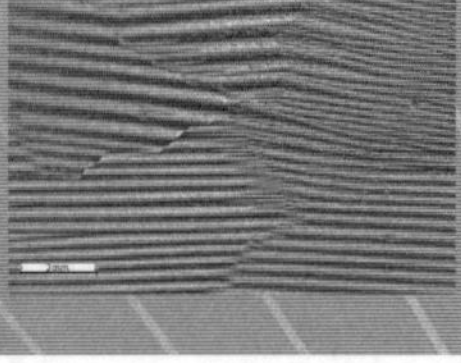

8

Magnetic field visualization: MO sensors

Physical background

- The underlying principle of magneto optics is the **Faraday effect**

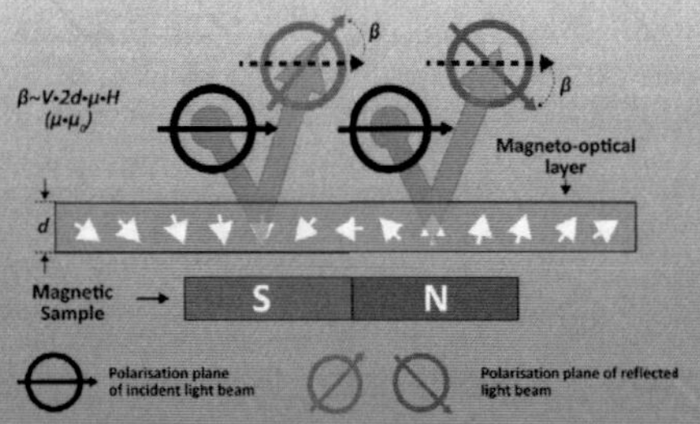

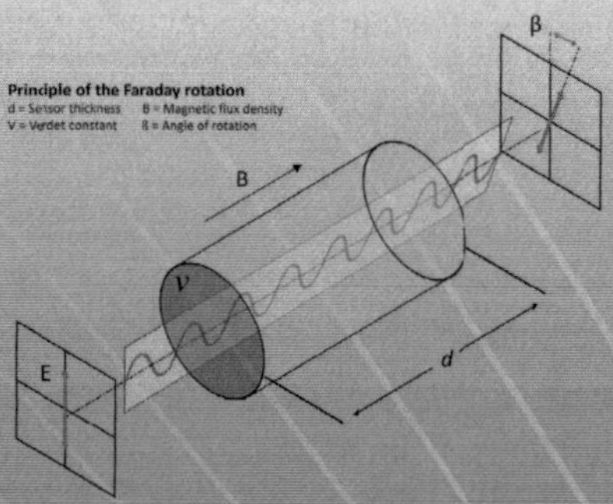

9

Magnetic field visualization: cmos-magview

Visualization of magnetic fields and magnetic structures

- **cmos-magview** is a high-resolution magnetic measurement and visualization system
- Use of the Faraday effect in the magneto-optical sensor
- Visualization of the magnetic stray fields near the sample surface
- Recording the magneto-optical image via high-resolution digital camera
- Analysis of polarity, homogeneity, distribution of magnetic material and stray field geometry in the micron range

10

Magnetic field visualization: cmos-magview

Visualization of magnetic fields and magnetic structures

- Available in 3 versions: S, L and XL
- Make magnetic field directly visible and measure magnetic flux density
- Measurement duration: 1 second
- Sensor size: up to 45 x 60 mm^2
- Geometric resolution: up to 25 µm (depending on sensor and camera)
- Software provides various views, measurement tools and settings

11

Magnetic field visualization: cmos-magview

Visualization of magnetic fields and magnetic structures

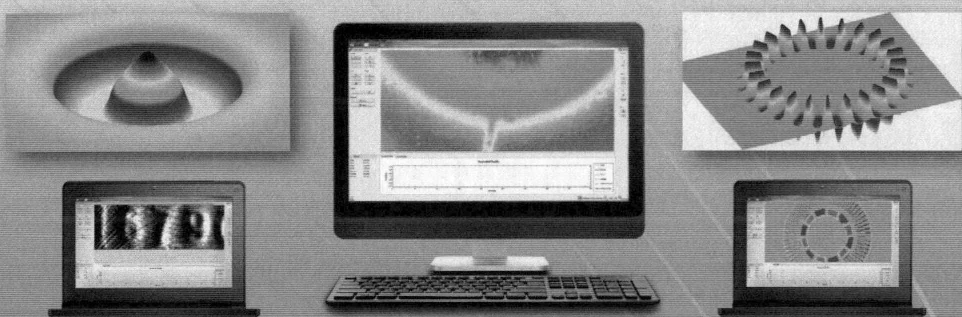

12

cmos-magview: automated image analysis

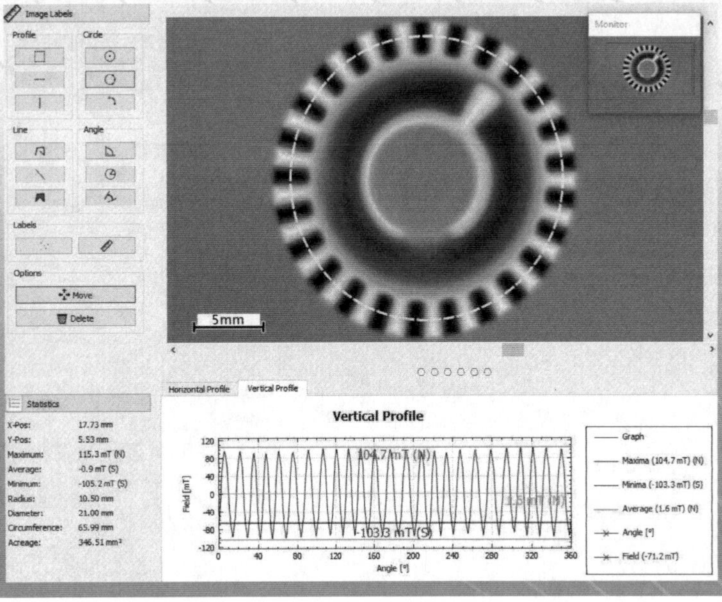

- Costumer specific software for automated detection of magnetic pole widths,
- Repeatability of angular measurements 0,02°
- Rotation encoders up to 38 mm diameter possible,
- Calibrated measurement range: ± 125 mT,
- Lateral resolution up to 60 µm with 45 mm x 60 mm sensor

13

Magnetic Measurement & Calibration Systems

Systems and technologies for (quality) control of permanent magnets and magnet systems

With our technologies, magnets can be :

- Characterized, classified, detected and tracked
- Generation and compensation of magnetic fields
- Fast and accurate quality inspection
- High-precision measurement
- Automation possible

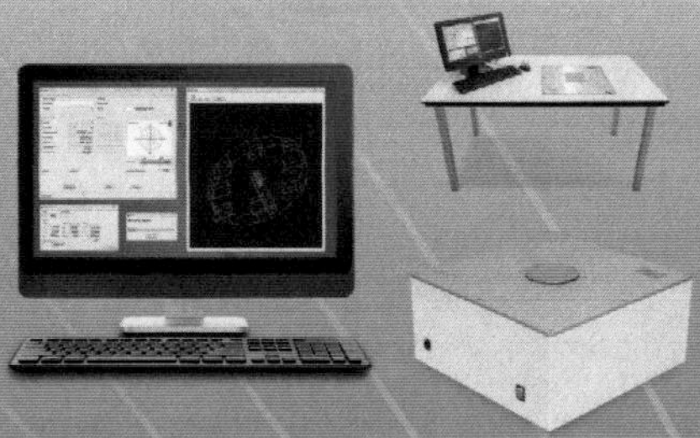

14

Magnetic Measurement & Calibration Systems: m-axis

Testing and measuring of permanent magnets

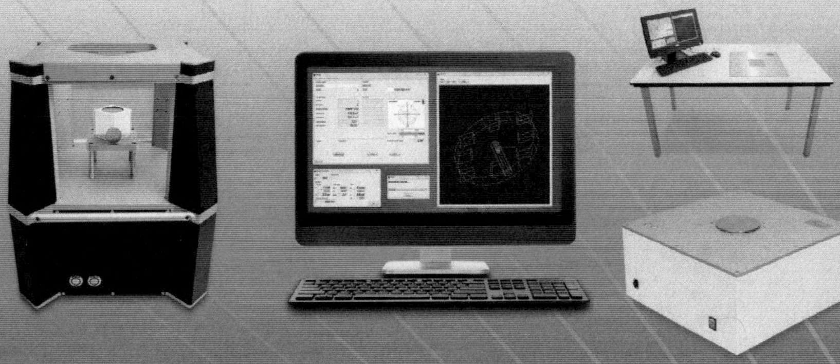

15

Magnetic Measurement & Calibration Systems: m-axis

- The m-axis is a measuring system with which permanent magnets can be characterized with high precision regarding their magnetic specifications (magnetic moment and magnetization angle)
- In addition to the 3 standard measuring devices (3 standard measuring ranges), there is also the option of adapting the m-axis for customer-specific measuring ranges for a 100% quality control

Standard Measurement Systems :

- m-axis S measuring range : 0.01 – 1.00 Am^2
- m-axis M measuring range : 0.10 – 3.00 Am^2
- m-axis L measuring range : 1.00 – 10.00 Am^2
- m-axis XL measuring range : 4.00 – 50.00 Am^2
- Customer-specific measuring ranges

16

Magnetic Measurement & Calibration Systems: m-axis

Highlights

- Determination of the magnetic moment
- Calculation of remanence
- Determination of the magnetization orientation
- Tabular and graphical representation of the size and orientation of the error angle
- Geometry independent characteristic determination
- Statistical evaluation of the parameters possible
- Automated storage of the determined characteristics in a CSV file
- Implementation of the system in an already existing manufacturing process possible

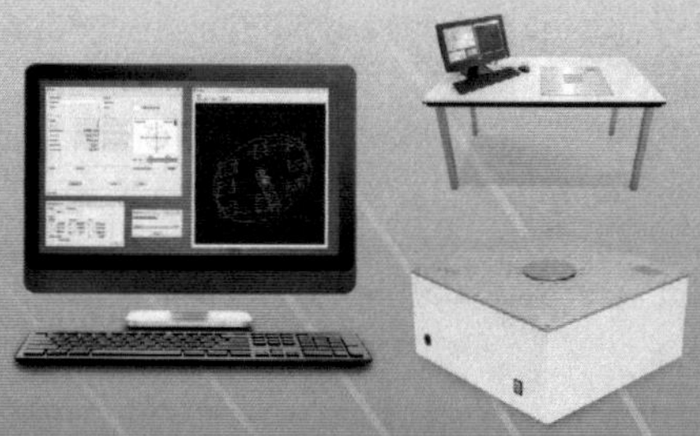

17

Magnetic Measurement & Calibration Systems: m-axis

Technical features

Accuracy moment and Br: ± 1%

- Repeatability moment and Br: ± 0.1%
- Accuracy magnetization angle: ± 0.1 °
- Fast classification
- North / south effect
- PC control with USB interface
- Mains connection: 85 - 264 VAC / 47 - 63 Hz
- Dimensions: 400 x 400 x 250 mm

18

Magnetic Measurement & Calibration Systems: m-axis

Optional modules

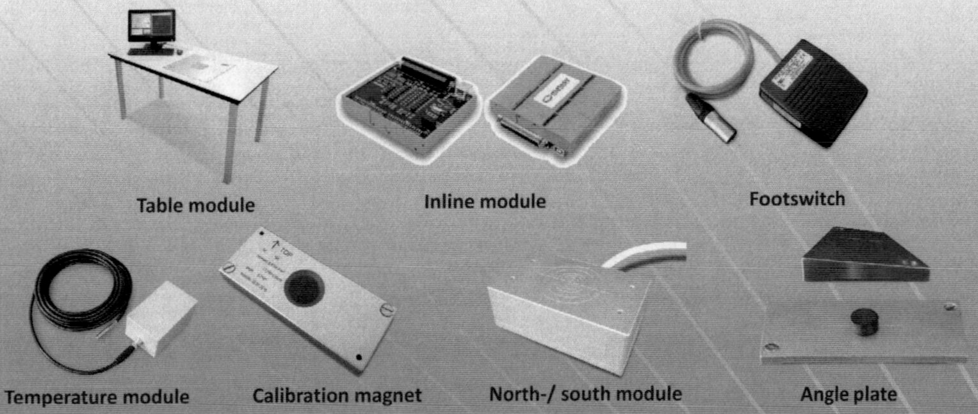

19

Magnetic Measurement & Calibration Systems: m-axis

Software:

- direct measurement of the magnetic moment and magnetization angle
- display of measurement quality during the process
- automatically generated protocols with all relevant settings and measurement data
- measurement processes are transferable to other m-axis systems

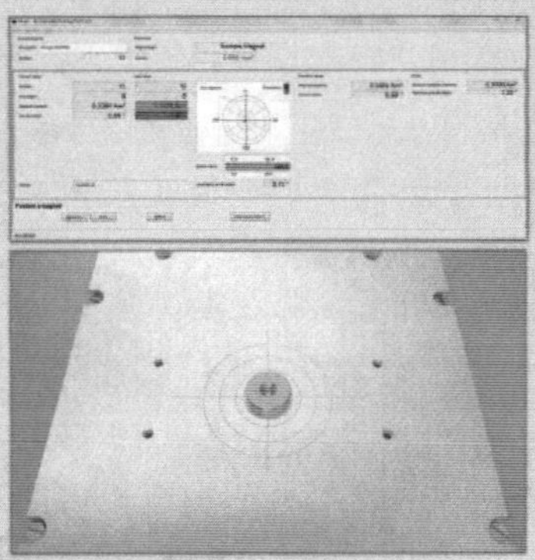

m-axis

20

m-axis I/O module for inline application

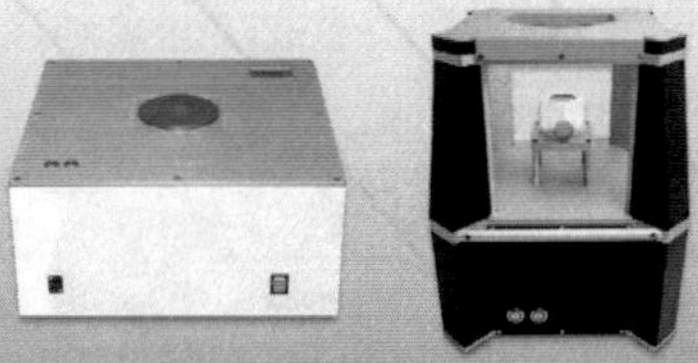

m-axis

I/O module

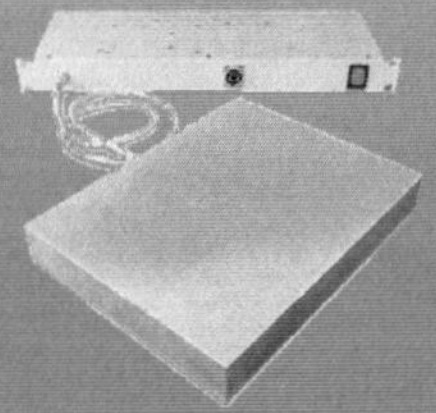

Automated quality control

21

Automated magnet measuring line in automotive industry

Measurement and alignment of the magnets

100% end-of-line test of the product

22

Automated magnet measuring line – magnet producer

23

Automated magnet measuring system – prototype Germany

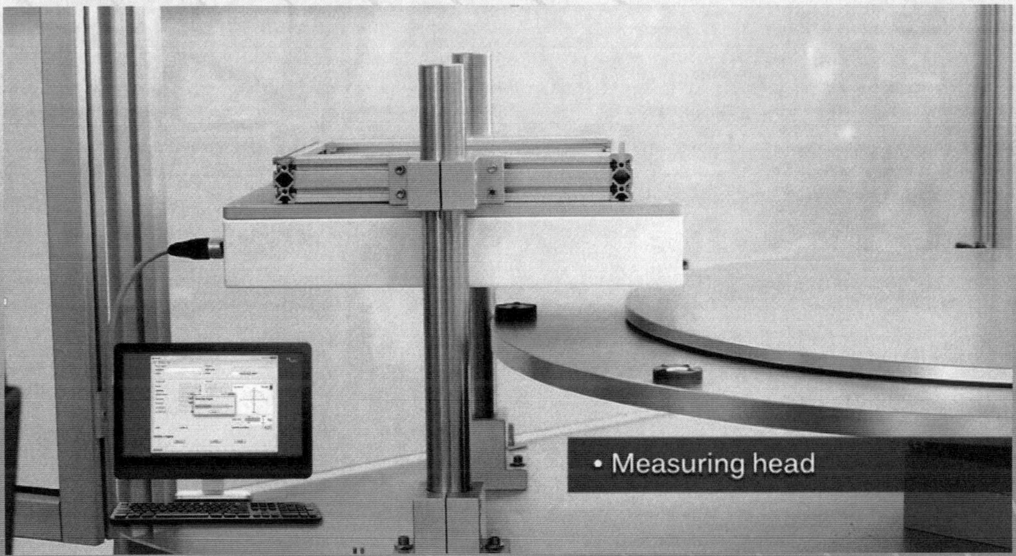

24

Magnetic Measurement : MHLS – Matesy Hall Line Sensor

Sensors:
- Chip size package (close to the surface)
- Real x, y and z Hall sensors
- x, y, z sensors in an area of 0.1 mm^3

Technical features:
- Up to 64 IC's (2 x 3 Hall Sensors inside)
- With 200Hz @16Bit
- Sensor pitch 2.5mm
- Up to 317.5 mm measuring length
- ± 800 mT measurement range
- Interface: USB, Serial (up to 10 m cable)
- Plug and play with software
- Trigger mode

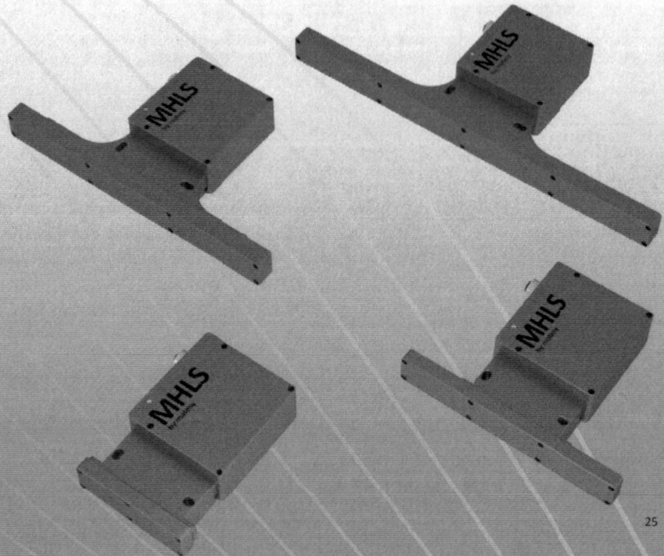

ZIM
Zentrales Innovationsprogramm Mittelstand
Gefördert durch:
Bundesministerium für Wirtschaft und Energie
aufgrund eines Beschlusses des Deutschen Bundestages

25

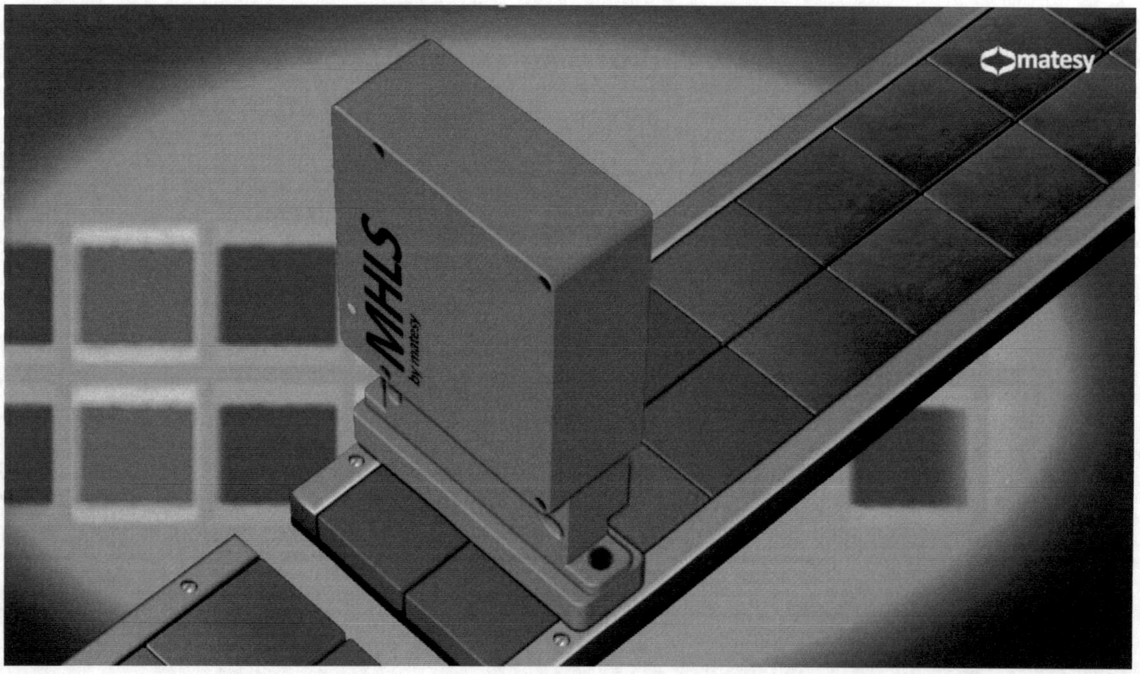

MHLS – Matesy Hall Line Sensor

- Easy integration to our rotor mapping system and industrial production lines
- Fast and precise measurements for 100% control
- Triggered measurements via hardware or software

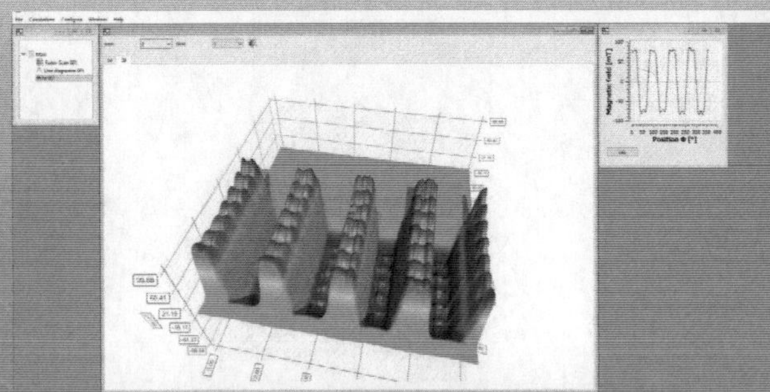

Conclusion

- **Image processing makes it possible to use magneto-optical systems for quality control in industrial production lines**
- **The inline module of the m-axis family enables the use of the system for 100% control of magnetic parameters of permanent magnets directly in the production**
- **The Matesy Hall Line Sensor enables rapid integration into existing industrial production lines or mappers for testing permanent magnet rotors and magnetic assemblies**

Thank you!

29

Address & Contact

matesy
Magnetic Technologies & Systems

Matesy GmbH

Löbstedter Str. 101

07749 Jena

Germany

Benjamin Wenzel

Tel: +49 (0) 3641 79799 00

Fax:+49 (0) 3641 79799 01

E-mail: b.wenzel@matesy.de

info@matesy.de / sales@matesy.de

Web: www.matesy.com

30

Four Steps to Minimize Costs and Maximize Profits

References

- Argus Media
- Asian Metal
- Oanda
- USGS
- US Department of Commerce
- Institute for Trend Research
- Several magnet manufacturers

First Derivative

- Given a function $f(x)$ the first derivative is defined as
- $f'(x) = \frac{f(x+h)-f(x)}{h}$
- In other words we find the value of a point on the curve of $f(x)$ and then the value of
- $f(x + h)$
- where h is another point on the curve. As h is allowed to go to zero, the quotient gives the instantaneous rate of change of $f(x) when\ x = a$. In similar fashion repeating the process using $f'(x)$, the second derivative $f''(x)$ is found.
- This is done because data can be analyzed.
- If the first derivative, $f'(x) > 0$, then $f(x)$ is increasing at $x = a$.
- If $f'(x) < 0$, then $f(x)$ is decreasing at $x = a$.
- The first derivative only reveals the slope, positive or negative, of the function at the point in question.

The Second Derivative

- The second derivative reveals the concavity of the curve.
- If the first derivative is positive the function is increasing. And if the second is positive, the curve is concave upward.
- If the first derivative is positive the function is increasing. And if the second is negative, the curve is concave down.
- If the first derivative is negative the function is decreasing. And the second is positive, the curve is concave upward.
- If the first derivative is negative the function is decreasing. And the second is negative, the curve is concave down.

Average Monthly Cost for Aluminum

USD / Kg	Al
Jan-17	1.594
Feb-17	1.634
Mar-17	1.702
Apr-17	1.691
May-17	1.671
Jun-17	1.621
Jul-17	1.642
Aug-17	1.741
Sep-17	1.771
Oct-17	1.834
Nov-17	1.862
Dec-17	1.819
Jan-18	1.834
Feb-18	1.841
Mar-18	1.848
Apr-18	1.893
May-18	1.898
Jun-18	1.757
Jul-18	1.764
Aug-18	1.648
Sep-18	1.544
Oct-18	1.437
Nov-18	1.385
Dec-18	1.411

Robert Wolf Data Decisions 219-309-5222 5

Aluminum Metal-Average Monthly Cost

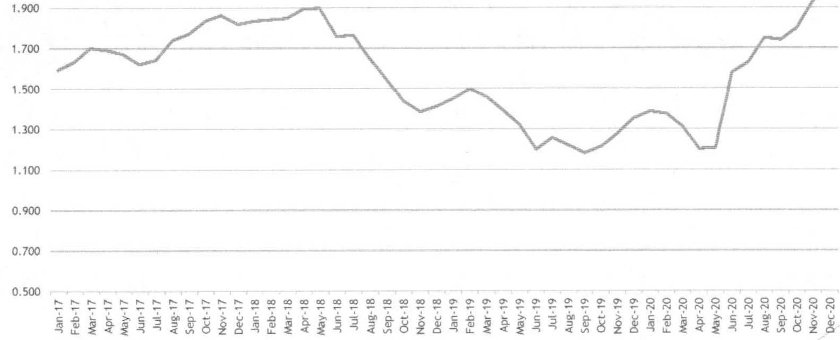

Robert Wolf Data Decisions 219-309-5222 6

Three Month Moving Total 3MMT

	Al	Al
USD / Kg		3MMT
Jan-17	1.594	
Feb-17	1.634	
Mar-17	1.702	4.930
Apr-17	1.691	5.027
May-17	1.671	5.064
Jun-17	1.621	4.983
Jul-17	1.642	4.934
Aug-17	1.741	5.004
Sep-17	1.771	5.154
Oct-17	1.834	5.346
Nov-17	1.862	5.467
Dec-17	1.819	5.515
Jan-18	1.834	5.514
Feb-18	1.841	5.493
Mar-18	1.848	5.522
Apr-18	1.893	5.582
May-18	1.898	5.638
Jun-18	1.757	5.548
Jul-18	1.764	5.419
Aug-18	1.648	5.169
Sep-18	1.544	4.956
Oct-18	1.437	4.629
Nov-18	1.385	4.367
Dec-18	1.411	4.234
Jan-19	1.450	4.247
Feb-19	1.499	4.361
Mar-19	1.459	4.408
Apr-19	1.390	4.348
May-19	1.321	4.170

Twelve Month Moving Total 12MMT

	Al	Al	Al
USD / Kg		3MMT	12MMT
Jan-17	1.594		
Feb-17	1.634		
Mar-17	1.702	4.930	
Apr-17	1.691	5.027	
May-17	1.671	5.064	
Jun-17	1.621	4.983	
Jul-17	1.642	4.934	
Aug-17	1.741	5.004	
Sep-17	1.771	5.154	
Oct-17	1.834	5.346	
Nov-17	1.862	5.467	
Dec-17	1.819	5.515	20.581
Jan-18	1.834	5.514	20.821
Feb-18	1.841	5.493	21.028
Mar-18	1.848	5.522	21.174
Apr-18	1.893	5.582	21.376
May-18	1.898	5.638	21.603
Jun-18	1.757	5.548	21.739
Jul-18	1.764	5.419	21.861
Aug-18	1.648	5.169	21.768
Sep-18	1.544	4.956	21.541
Oct-18	1.437	4.629	21.145
Nov-18	1.385	4.367	20.668
Dec-18	1.411	4.234	20.260
Jan-19	1.450	4.247	19.877
Feb-19	1.499	4.361	19.535
Mar-19	1.459	4.408	19.146

3MMT Compared to the 3MMT 12 Months Earlier

	Al	Al	Al	Al
USD / Kg		3MMT	12MMT	3-12MMT
Jan-17	1.594			
Feb-17	1.634			
Mar-17	1.702	4.930		
Apr-17	1.691	5.027		
May-17	1.671	5.064		
Jun-17	1.621	4.983		
Jul-17	1.642	4.934		
Aug-17	1.741	5.004		
Sep-17	1.771	5.154		
Oct-17	1.834	5.346		
Nov-17	1.862	5.467		
Dec-17	1.819	5.515	20.581	
Jan-18	1.834	5.514	20.821	
Feb-18	1.841	5.493	21.028	
Mar-18	1.848	5.522	21.174	12.0%
Apr-18	1.893	5.582	21.376	11.0%
May-18	1.898	5.638	21.603	11.3%
Jun-18	1.757	5.548	21.739	11.3%
Jul-18	1.764	5.419	21.861	9.8%
Aug-18	1.648	5.169	21.768	3.3%
Sep-18	1.544	4.956	21.541	-3.8%
Oct-18	1.437	4.629	21.145	-13.4%
Nov-18	1.385	4.367	20.668	-20.1%
Dec-18	1.411	4.234	20.260	-23.2%
Jan-19	1.450	4.247	19.877	-23.0%
Feb-19	1.499	4.361	19.535	-20.6%
Mar-19	1.459	4.408	19.146	-20.2%
Apr-19	1.390	4.348	18.643	-22.1%
May-19	1.321	4.170	18.067	-26.0%

Robert Wolf Data Decisions 219-309-5222 9

12 MMT Compared to the 12MMT 12 Months Earlier

	Al	Al	Al	Al	Al
USD / Kg		3MMT	12MMT	3-12MMT	12-12 MMT
Jan-17	1.594				
Feb-17	1.634				
Mar-17	1.702	4.930			
Apr-17	1.691	5.027			
May-17	1.671	5.064			
Jun-17	1.621	4.983			
Jul-17	1.642	4.934			
Aug-17	1.741	5.004			
Sep-17	1.771	5.154			
Oct-17	1.834	5.346			
Nov-17	1.862	5.467			
Dec-17	1.819	5.515	20.581		
Jan-18	1.834	5.514	20.821		
Feb-18	1.841	5.493	21.028		
Mar-18	1.848	5.522	21.174	12.0%	
Apr-18	1.893	5.582	21.376	11.0%	
May-18	1.898	5.638	21.603	11.3%	
Jun-18	1.757	5.548	21.739	11.3%	
Jul-18	1.764	5.419	21.861	9.8%	
Aug-18	1.648	5.169	21.768	3.3%	
Sep-18	1.544	4.956	21.541	-3.8%	
Oct-18	1.437	4.629	21.145	-13.4%	
Nov-18	1.385	4.367	20.668	-20.1%	
Dec-18	1.411	4.234	20.260	-23.2%	-1.6%
Jan-19	1.450	4.247	19.877	-23.0%	-4.5%
Feb-19	1.499	4.361	19.535	-20.6%	-7.1%
Mar-19	1.459	4.408	19.146	-20.2%	-9.6%
Apr-19	1.390	4.348	18.643	-22.1%	-12.8%
May-19	1.321	4.170	18.067	-26.0%	-16.4%

Robert Wolf Data Decisions 219-309-5222 10

Aluminum Average Monthly Costs

Robert Wolf Data Decisions 219-309-5222 11

Aluminum Metal

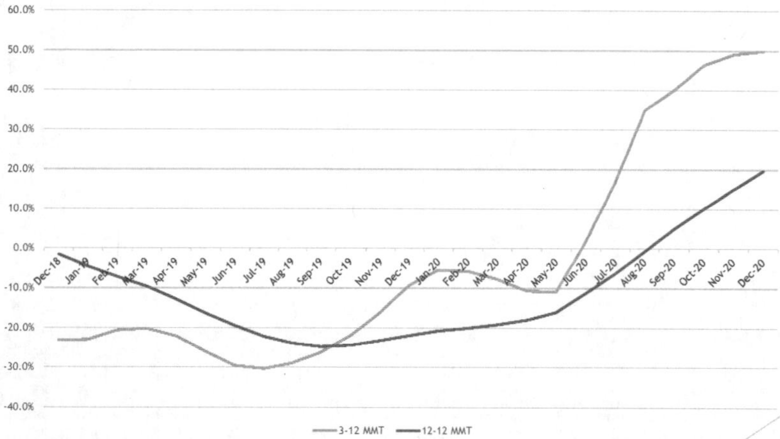

Robert Wolf Data Decisions 219-309-5222 12

Nickel Metal

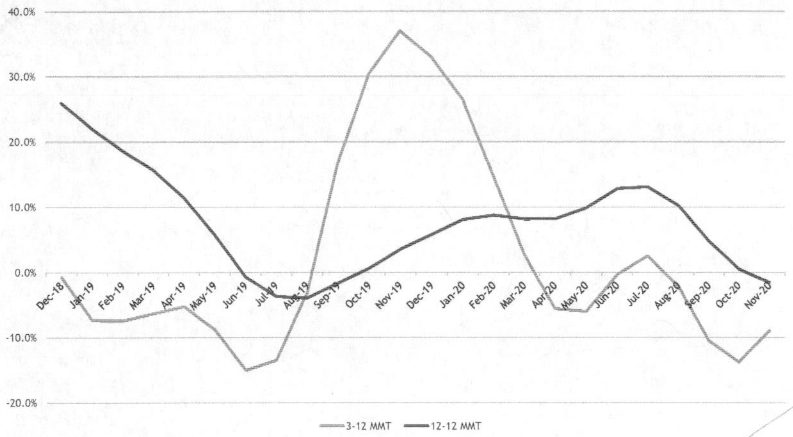

Robert Wolf Data Decisions 219-309-5222 13

Cobalt Metal

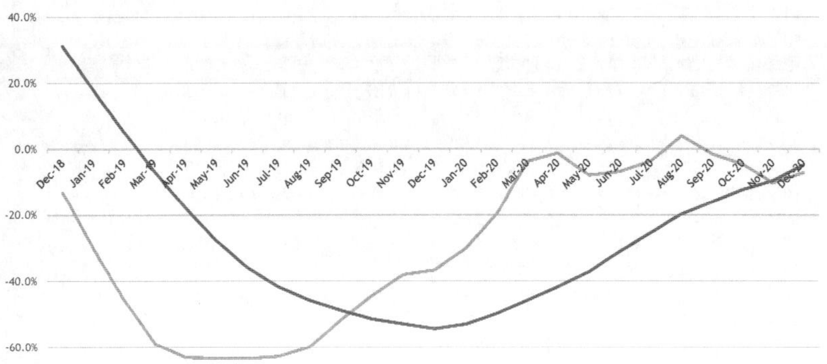

Robert Wolf Data Decisions 219-309-5222 14

Copper Metal

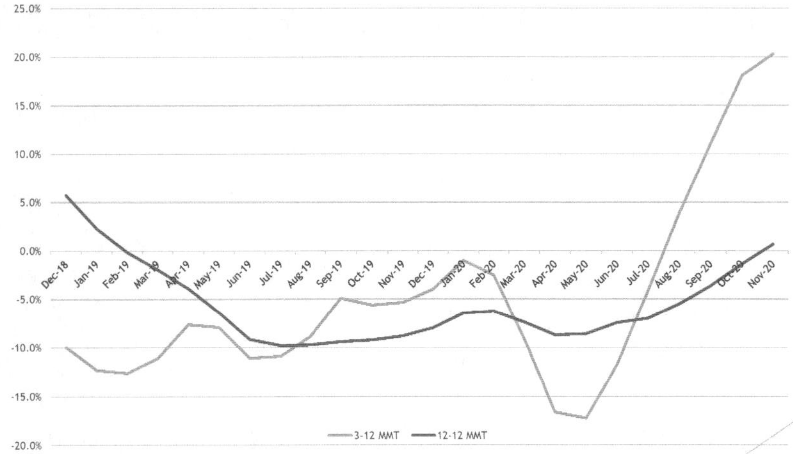

Robert Wolf Data Decisions 219-309-5222 15

Average Metal Price per Month ($/Kg)

	Al	Ni	Co	Cu	Sm	Nd	Fe	FeB	Dy
USD / Kg									
Jan-17	1.594	10.033	34.724	5.758	14.700	49.800	0.080	3.190	261.000
Feb-17	1.634	10.661	42.931	5.960	14.750	50.410	0.089	2.915	261.000
Mar-17	1.702	10.280	52.805	5.837	14.685	51.260	0.088	3.091	261.000
Apr-17	1.691	9.723	55.316	5.725	14.450	51.755	0.070	3.355	261.000
May-17	1.671	9.190	54.661	5.614	14.450	53.390	0.062	3.355	261.000
Jun-17	1.621	8.982	57.406	5.724	14.450	55.075	0.057	3.355	261.000
Jul-17	1.642	9.555	57.451	6.018	14.525	61.875	0.068	3.223	266.125
Aug-17	1.741	10.893	57.650	6.504	14.750	87.400	0.076	3.091	282.600
Sep-17	1.771	11.269	60.008	6.612	14.845	95.000	0.072	3.091	284.380
Oct-17	1.834	11.378	59.545	6.846	14.900	83.500	0.062	3.047	277.500
Nov-17	1.862	12.060	61.702	6.846	14.900	72.500	0.064	2.849	277.500
Dec-17	1.819	11.491	72.453	6.826	14.900	60.570	0.072	2.893	266.430
Jan-18	1.834	12.914	77.560	7.118	15.035	66.275	0.076	2.937	270.670
Feb-18	1.841	13.593	81.388	7.038	15.300	67.850	0.077	3.069	274.000
Mar-18	1.848	13.429	87.940	6.831	15.380	73.250	0.070	3.135	279.330
Apr-18	1.893	13.931	90.917	6.870	16.000	70.130	0.066	3.300	279.750

Robert Wolf Data Decisions 219-309-5222 16

Typical Magnet Chemistry

Magnet Grade	Alnico 5	Alnico 8	SmCo	NdFeB	NdFeB-H	NdFeB-SH	NdFeB-UH	NdFeB-EH
Element								
Al	8.00%	7.00%	0.00%	0.00%	0.00%	0.00%	0.00%	0.00%
Ni	14.00%	15.00%	0.00%	0.00%	0.00%	0.00%	0.00%	0.00%
Co	24.00%	35.00%	65.00%	0.00%	1.00%	1.20%	2.00%	3.00%
Cu	3.00%	4.00%	0.00%	0.00%	0.00%	0.00%	0.00%	0.00%
Sm	0.00%	0.00%	35.00%	0.00%	0.00%	0.00%	0.00%	0.00%
Nd	0.00%	0.00%	0.00%	34.00%	33.00%	31.00%	31.00%	31.00%
Fe	51.00%	39.00%	0.00%	65.00%	63.00%	63.00%	61.50%	60.00%
FeB	0.00%	0.00%	0.00%	1.00%	1.00%	1.00%	1.00%	1.00%
Dy	0.00%	0.00%	0.00%	0.00%	2.00%	3.80%	4.50%	5.00%

Average Monthly Cost for Magnet Material

	Alnico 5	Alnico 8	SmCo	NdFeB	NdFeB-H	NdFeB-SH	NdFeB-UH	NdFeB-EH
USD / Kg								
Jan-17	$10.079	$14.031	$27.715	$17.016	$22.084	$25.855	$27.959	$29.610
Feb-17	$12.151	$17.013	$33.068	$17.226	$22.370	$26.145	$28.315	$30.048
Mar-17	$14.469	$20.411	$39.463	$17.517	$22.750	$26.529	$28.777	$30.608
Apr-17	$14.980	$21.194	$41.013	$17.676	$22.930	$26.703	$28.972	$30.829
May-17	$14.739	$20.876	$40.587	$18.226	$23.458	$27.197	$29.461	$31.311
Jun-17	$15.365	$21.804	$42.371	$18.796	$24.038	$27.750	$30.035	$31.913
Jul-17	$15.473	$21.923	$42.427	$21.114	$26.391	$30.058	$32.380	$34.284
Aug-17	$15.734	$22.223	$42.635	$29.796	$35.149	$38.603	$41.042	$43.030
Sep-17	$16.356	$23.110	$44.201	$32.378	$37.714	$41.053	$43.522	$45.543
Oct-17	$16.267	$22.974	$43.920	$28.461	$33.770	$37.214	$39.632	$41.614
Nov-17	$16.884	$23.834	$45.322	$24.720	$30.161	$33.829	$36.264	$38.268
Dec-17	$19.384	$27.510	$52.309	$20.670	$26.116	$29.845	$32.288	$34.344
Jan-18	$20.821	$29.526	$55.676	$22.612	$28.137	$31.839	$34.353	$36.481
Feb-18	$21.834	$30.965	$58.257	$23.150	$28.764	$32.501	$35.069	$37.252
Mar-18	$23.374	$33.223	$62.544	$24.982	$30.714	$34.453	$37.111	$39.386
Apr-18	$24.162	$34.344	$64.696	$23.920	$29.722	$33.536	$36.221	$38.528
May-18	$24.056	$34.164	$64.197	$22.244	$28.073	$31.971	$34.645	$36.941
Jun-18	$22.090	$31.263	$58.328	$22.373	$28.127	$32.014	$34.624	$36.838

Alnico 5 Material Costs

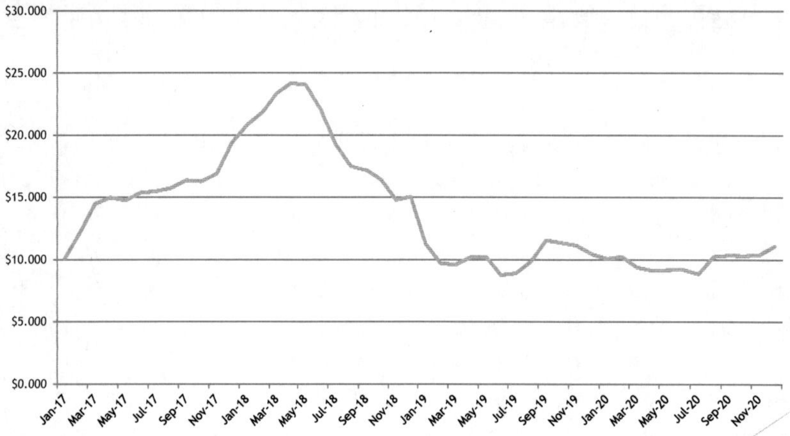

Robert Wolf Data Decisions 219-309-5222 19

Alnico 5 Material Costs

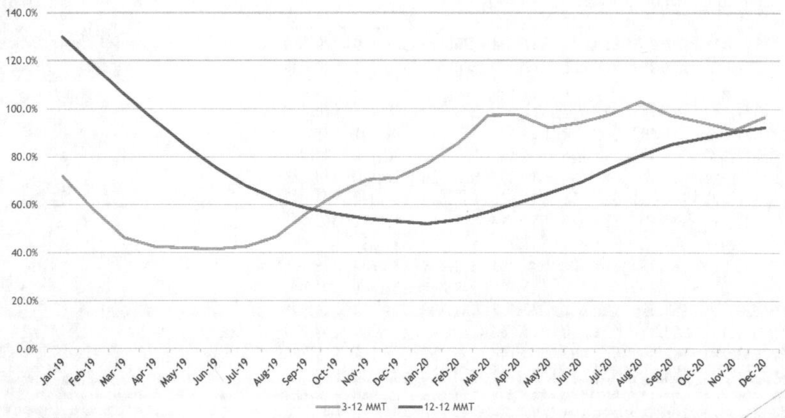

Robert Wolf Data Decisions 219-309-5222 20

NdFeB-SH Material Costs

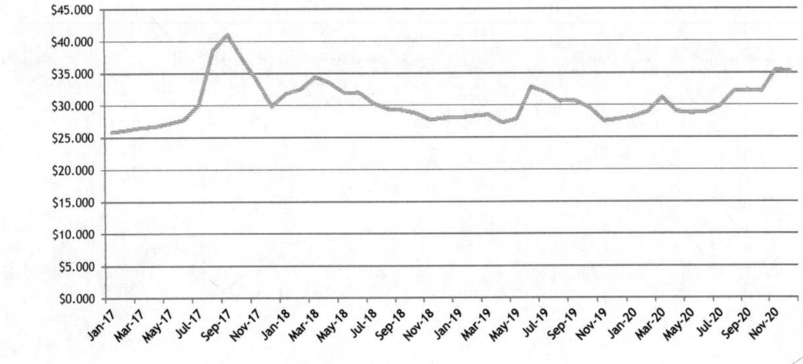

Robert Wolf Data Decisions 219-309-5222 21

NdFeB-SH Material Costs

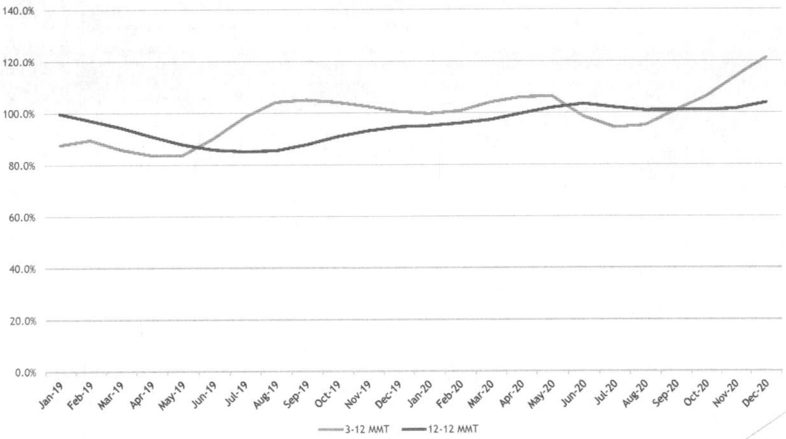

Robert Wolf Data Decisions 219-309-5222 22

Comment

- Specific types and grades are not represented on this list nor are exact chemistries as these are proprietary to the individual magnet manufacturers. If more details are desired, consult with your specific manufacturer
- Treating the data on the chemistries as matrix A and the monthly prices of the elements as matrix B, it is possible to multiply the two matrices and arrive at the average monthly price for the different types of magnets. CAUTION you only have the cost for the elements exclusive of any manufacturing processes, SGA, taxes and duties and transportation. These can be accounted for in additional matices or multipliers.

ADDITIVE MANUFACTURING OF MAGNETIC MATERIALS

Dr. Sim Narasimhan

P2PTechnologies

Moorestown ,New Jersey

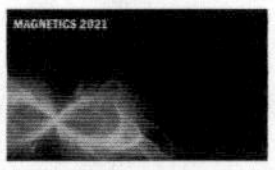

P2P TECHNOLOGIES
Magnets,Powdermetal

P2P Technologies is a Division of CS 21 LLC, registered in NJ, USA.

Intellectual Property strategies; Business Strategies

Guidance in material selection and processing of metal powders

Sales of Neo , rare earth,   permanent magnets

Global Sales Marketing including India, China, Korea and USA

Seminars on Powder Metallurgy, Permanent magnets for beginners and advanced level

OUTLINE OF THE PRESENTATION

- Brief intro to AM(3D) manufacturing
- Application to soft magnetic materials
- Update on AM for Permanent magnets
- Conclusions

MAGNETICS 2021

ADDITIVE MANUFACTURING OR 3DPRINTING

Method of building a part layer by layer using heat source or binder using a CAD software

Heat source generally used is laser or electron beam or plasma

ADDITIVE MANUFACTURING OR 3DPRINTING

Binder jetting, Directed energy deposition, Material extrusion, Material jetting, Powder bed fusion, Sheet lamination, Vat photo polymerization. Directed energy deposition which includes laser deposition (LD), laser induced net shaping (LENS), electron beam melting, plasma arc melting .

The process of 3D printing starts with 3D Modelling which calculates the amount of material to be deposited or melted in each layer, slicing the desired part so each layer deposited meets the part configuration . Identifying the machine to be used and the process parameters. Post processing involves machining, electrical discharge machining, peening, grinding, heat treatment, hot isostatic pressing.

The limitation on 3Dprinting is that part size is limited to powder bed size although direct deposition can make larger sizes but build times are slow. Production rate is limited about 25000/year. Non weldable materials cannot be processed. Densities in excess of 99.5% can be achieved. There is anisotropy in the build direction.

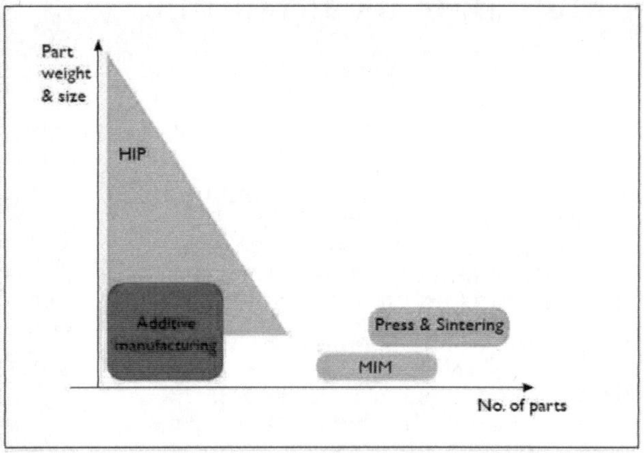

Positioning of various PM technologies according to part weight or size and production series

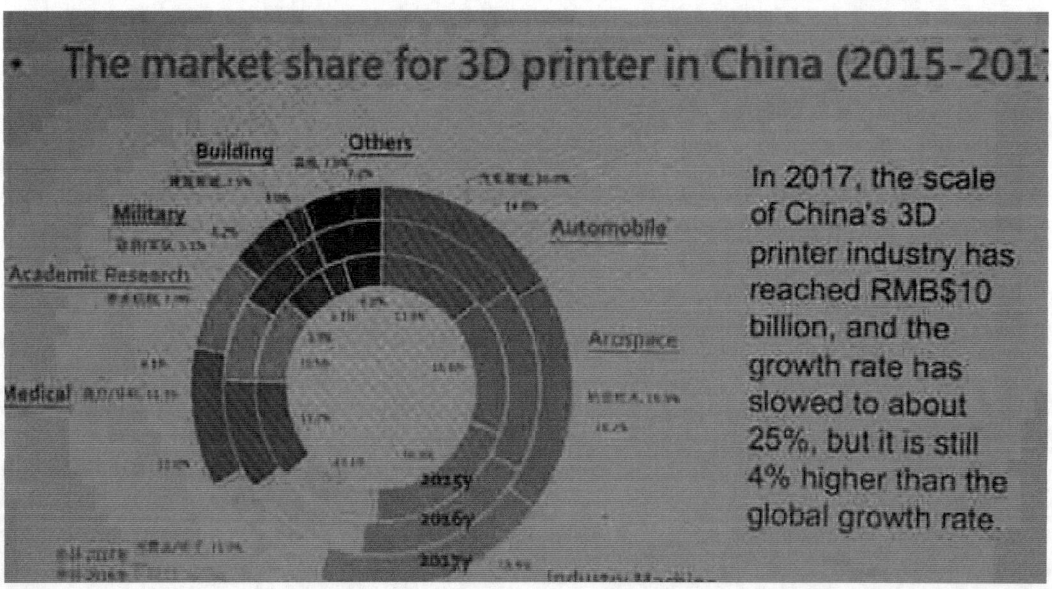

Complex shapes are possible with AM

Prototype of heat exchanger for electrical motor with struts of 0.4 mm diameter for COMPOLIGHT project (Courtesy of Fraunhofer IFAM)

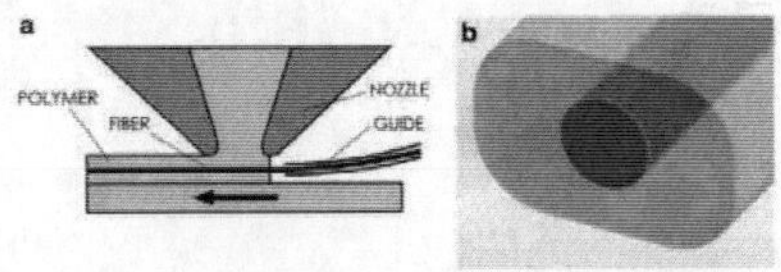

(a) schematic of FEAM printhead (b) thermoplastic and copper wire

www.nafems.org

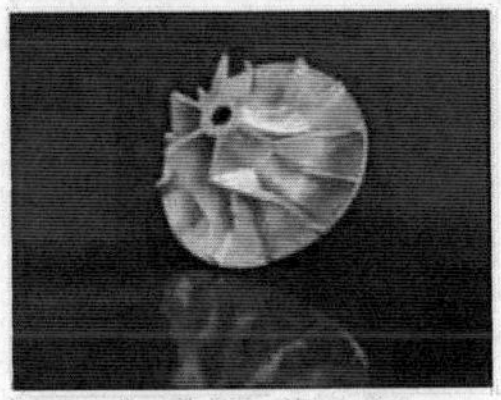

Turbine wheel by Metal Binder Jetting process (Courtesy of Fraunhofer IFAM)

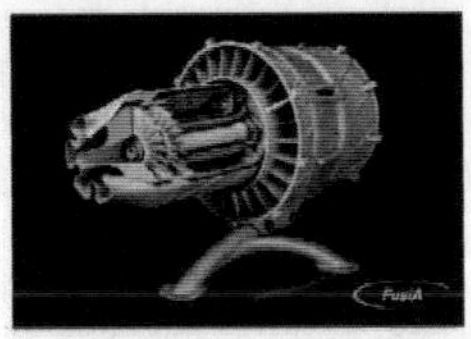

Gas turbine demonstrator (diameter 250 mm and length 600 mm), by assembling parts made by Selective Laser Melting with Al-, Ti- and Ni-base powders for integration of functions, reduced number of parts, weight saving and increase of performance

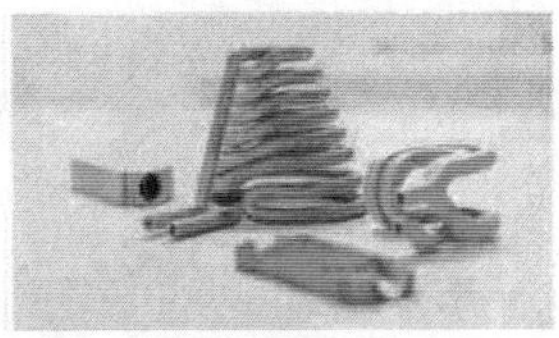

3D printed copper components by TRUMPF

The motor was produced using a method in which highly viscous metallic and ceramic pastes were extruded through a nozzle to build the body of the part in layers, before sintering to the required hardness, in a process the team refers to as 'Multimedia 3D Printing'.

stator for a winding-less reluctance machine produced by the researchers. Left: CAD model, middle: sintered stator, right: green part before heat treatment (Courtesy Johannes Rudolph)

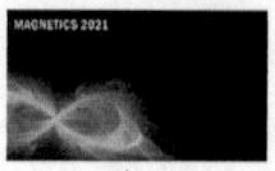

WHY 3D FOR MAGNETIC MATERIALS

- Design freedom compared to Press and sinter ;casting and machining
- Provides material where needed
- Net Shape capability saves raw material costs
- Magnets usage is based on ease of forming the Magnets, simple shapes
- Often design is compromised to fit magnet shapes mostly two dimensional
- Additive manufacturing ,3Dprinting allows complex shapes ,in 3D
- The technology allows possibility of generating fine nano size grains in bulk solid ,provide improved properties.

CURRENT TRENDS

- Growth of hybrid vehicles
- Permanent magnets increasing usage
- Process simplification and HSE
- Energy efficient motors
- Reduced weight in packaging electronic products
- Sensor usage is increasing
- Move from mechanical to electrical systems

SELECTIVE LASER MELTING

Binder Jetting

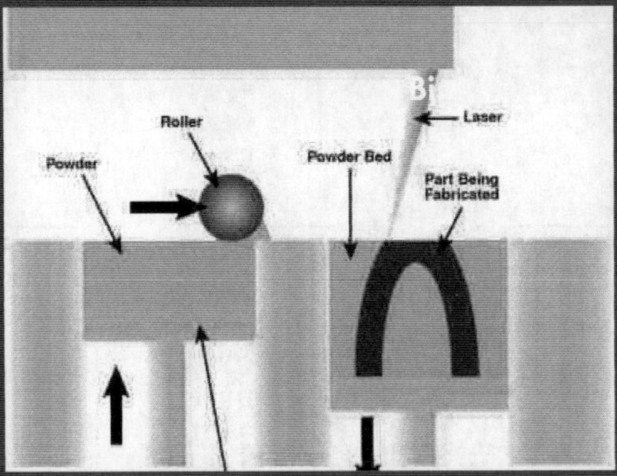

Small area is melted and Cooling is rapid and forms fine grains

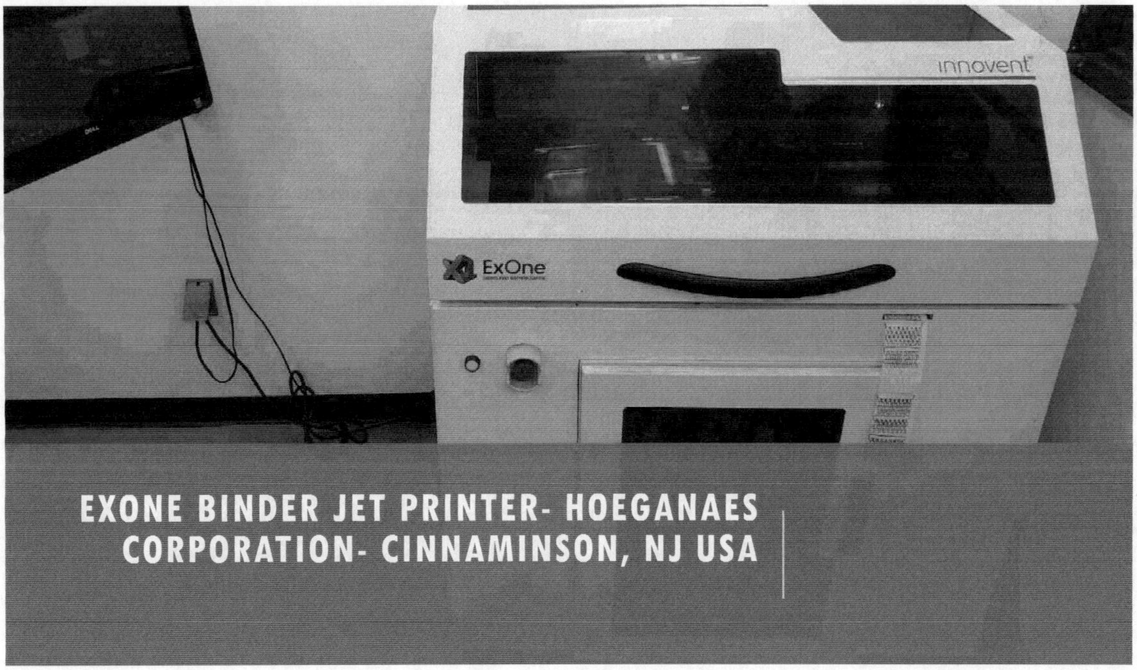

EXONE BINDER JET PRINTER- HOEGANAES CORPORATION- CINNAMINSON, NJ USA

EXONE BINDER JET PRINTER- HOEGANAES CORPORATION- CINNAMINSON, NJ USA

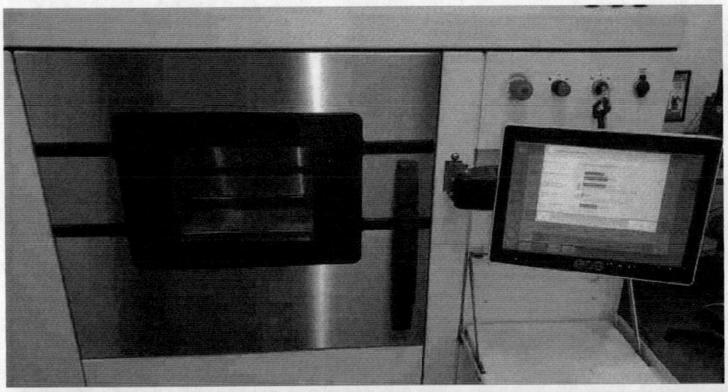

**EOS M290- HOEGANAES CORPORATION-
CINNAMINSON, NJ USA**

EOS M290- HOEGANAES CORPORATION- CINNAMINSON, NJ USA

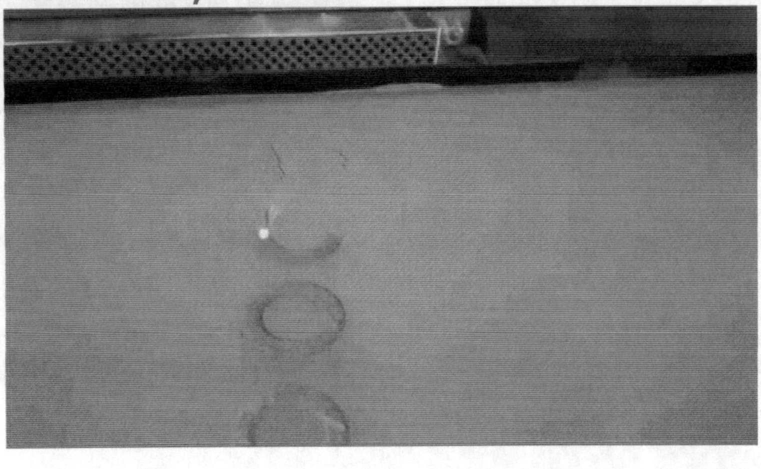

$$E = \frac{P}{v \cdot h \cdot t}$$

E = energy density

P = power (W)

v = scanning speed (mm/s)

h = hatch spacing (mm)

t = layer thickness (mm)

SOFT MAGNETIC MATERIALS

Soft magnetic materials are characterized by very low coercive force and a rapid magnetizing and demagnetizing cycles with low hysteresis losses.

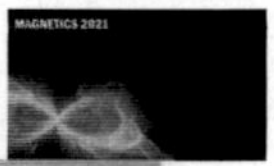

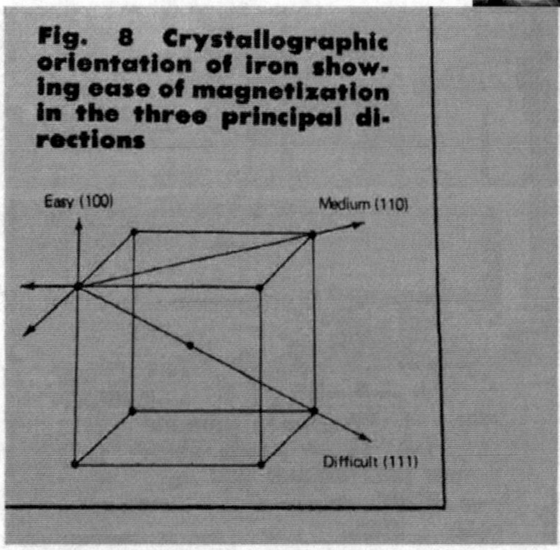

Fig. 8 Crystallographic orientation of iron showing ease of magnetization in the three principal directions

[redacted]

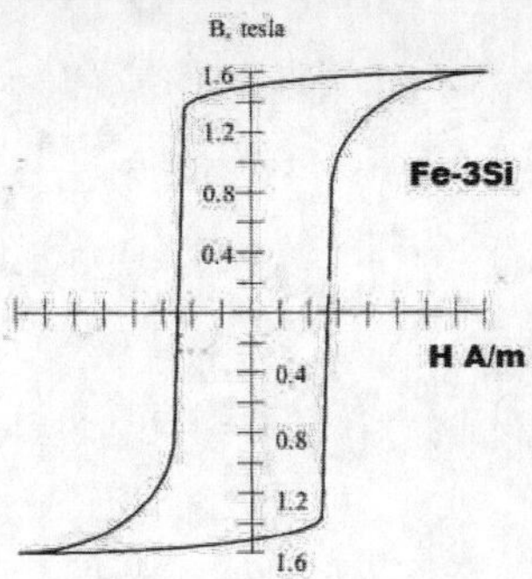

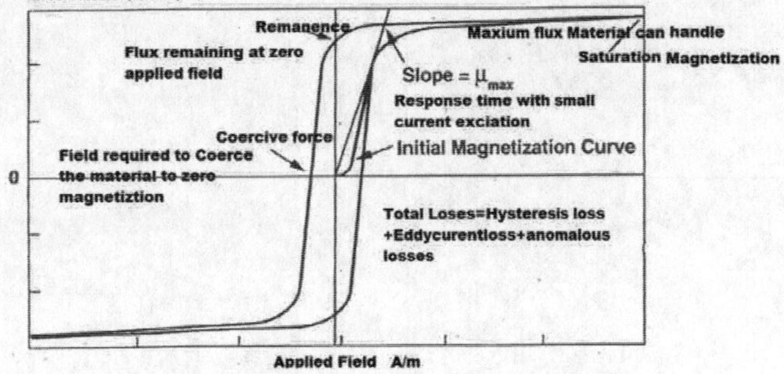

Soft Magnetic Materials-select

Grade	Composition	Saturation Induction KGauss(Tesla)	Coercive force A/m/(Oe)	Maxmium permeability	Resistivity Micro-Ohm cm
Fe-Si	1%Si	21000	35.2(0.442)	14780	16
	3.0Si	20500	15.2(0.192)	17440	45
	4.8Si	19300	15.9(0.20)	17160	60
	6.4Si	18200	11.7(0.147)	28860	70
DeltaMax	50Ni,balFe	16,000(1.6T)	3.18-12.75(0.04to0.16)	100000	45
Alloy4750	48Ni,Bal Fe	15,500(1.55T)	1.6-7.9(,012-,10)	100000	45
Mu metal	77Ni,5Cu,2Cr,0r4%Mo,balFe	7500(.75T)	0.8-2.4(.01-.03)	100000	60
Supermendur	49Co,2V,balFe	24000(2.4T)	16(0.2)	92000	40
Finemet	1Cu,3Nb,13.5Si,9B,73.5Fe	12400(1.24T)	0.5(0.06)	10000	~50
2714A	85Co,3Ni,	5700(0.57T)	---	100000	142
2826B	38Ni,4Mo,18B,40Fe	---	---	10000	142

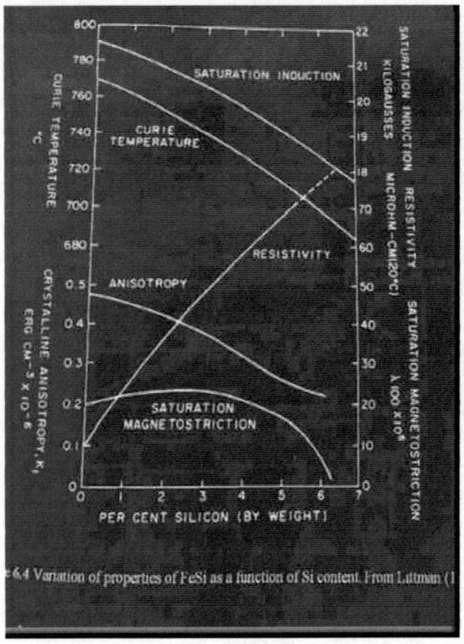
SATURATION INDUCTION
CURIE TEMPERATURE
RESISTIVITY
ANISOTROPY
SATURATION MAGNETOSTRICTION
CURIE TEMPERATURE °C
CRYSTALLINE ANISOTROPY, K_1 ERG CM^{-3} X 10^{-6}
SATURATION INDUCTION KILOGAUSSES
RESISTIVITY MICROHM-CM(20°C)
SATURATION MAGNETOSTRICTION λ100 X10^6
PER CENT SILICON (BY WEIGHT)
e 6.4 Variation of properties of FeSi as a function of Si content. From Littman (1

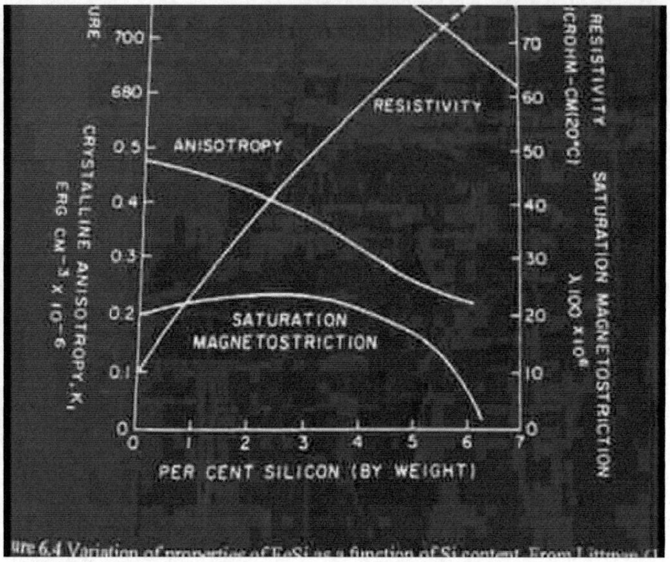
RESISTIVITY
ANISOTROPY
SATURATION MAGNETOSTRICTION
CRYSTALLINE ANISOTROPY, K_1 ERG CM^{-3} X 10^{-6}
RESISTIVITY
ICROHM-CM(20°C)
SATURATION MAGNETOSTRICTION λ100 X10^6
PER CENT SILICON (BY WEIGHT)

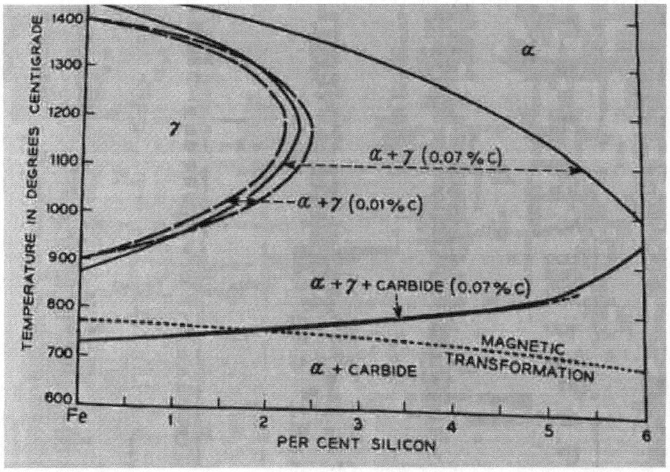
TEMPERATURE IN DEGREES CENTIGRADE
α
γ
α + γ (0.07 % C)
α + γ (0.01% C)
α + γ + CARBIDE (0.07 % C)
MAGNETIC TRANSFORMATION
α + CARBIDE
Fe
PER CENT SILICON

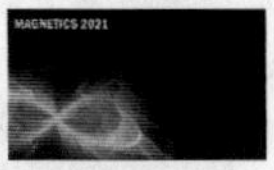

LBPF-3.5SI

Gas atomized 3.5% Si-Fe powder, 15-53 um

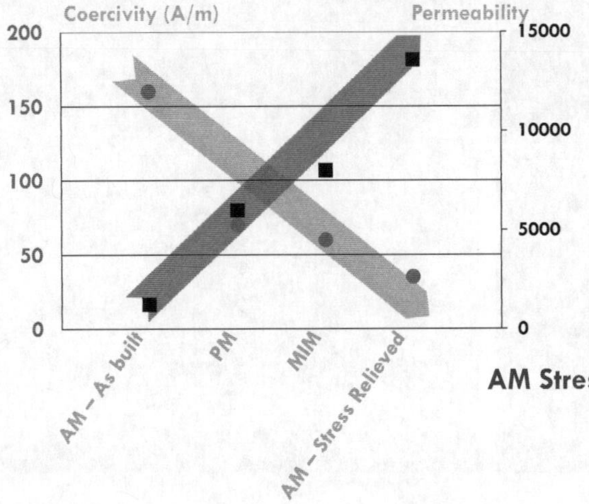

AM Stress Relieved – 7.6 g/cm³

Data form Hoeganaes Corporation

LBPF, Gas atomized 3.5% Si-Fe powder, 15-53 microns

H Field Max	Magnetic Flux Density (T)	Coercive Force (A/m)	Permeability
1,200 A/m	1.47	32	13680
5,000 A/m	1.66	32	13655
10,000 A/m	1.81	32	13595

FE-SILICON STEELS

Small transformers use 0.47mm lamination steels (M27, M36, M43 grades) and 0.64mm lamination steels (M45, M47)

Ballasts and Ignition Coils use 0.47 and 0.64 mm Lamination Steels (M27, M36, M43, M45, M47 Grade

Brushless DC Motors .

50 and 60Hz Induction Motors and Household Motors

Use Thinner laminations to reduce eddy current losses

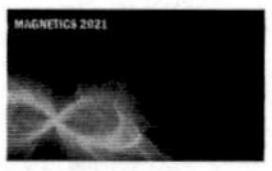

TARGET MARKET HIGH FREQUENCY MOTORS

Higher frequency gives higher efficiency, but also results in greater heat generation because of eddy currents generated in the part.

These eddy currents reduces the skin depth of material being magnetized during AC fields

We need to worry about eddy currents and skin depth

At a given flux density larger ferrite cores are needed- opposite to smaller package requirement in automotive.

Added core area increase copper losses

$$\delta = \sqrt{\frac{2\rho}{\omega\mu}}$$

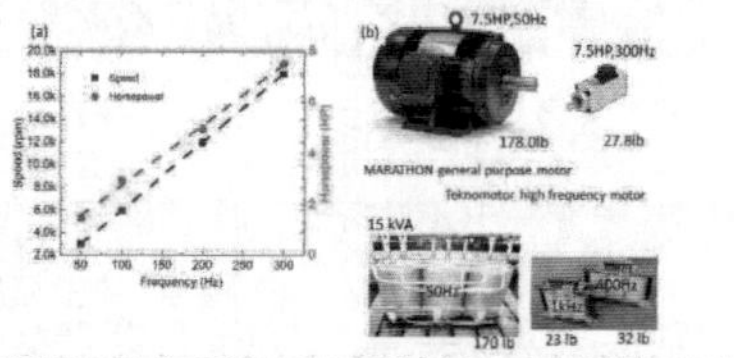

This core has been used in the TOYOTA PRIUS since 2009.

- Contains an inverter that produces 3-phase ac for driving the motor
- A dc-to-dc converter that produces 12V for auxiliary equipment
- Bi-directional, dc-to-dc converter that boosts the 200V battery up to a maximum of 500Vdc link,
- Or bucks the higher voltage dc link (that may range from 200–500V) down to charge the 200V battery.

The eddy current loss per unit volume is ;

- Proportional to the square of the frequency
- Proportional to the square of the flux density
- Proportional to the square of the thickness of the laminated sheet
- Inversely proportional to the resistivity of the material

$$P = \left(\frac{\pi^2}{6\rho}\right)(f^2\, B_{max}^2\, t^2)$$

Thickness can be reduced by using thin laminations or using powder metal coated with insulation

Resistivity can be improved by using higher silicon alloys

Induction B can be reduced by alloying

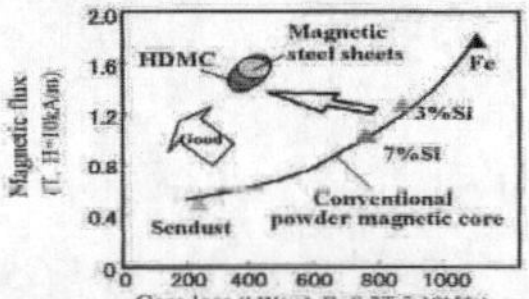

Magnetic properties

IP

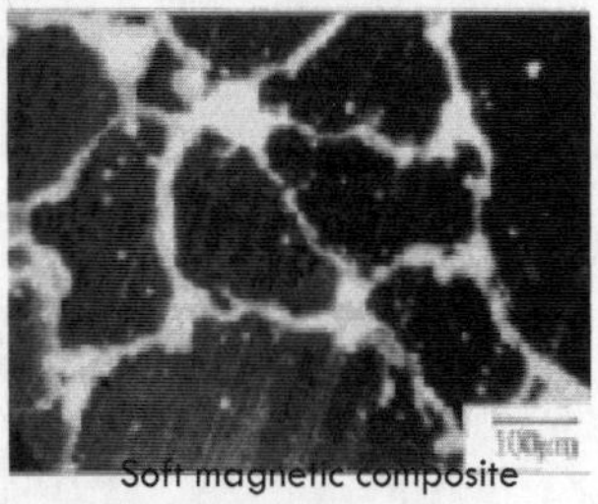

Soft magnetic composite

Hoeganaes Corporation Invented in 1990 thin coatings

METHODS OF MAKING FE-6.5SI STRIP

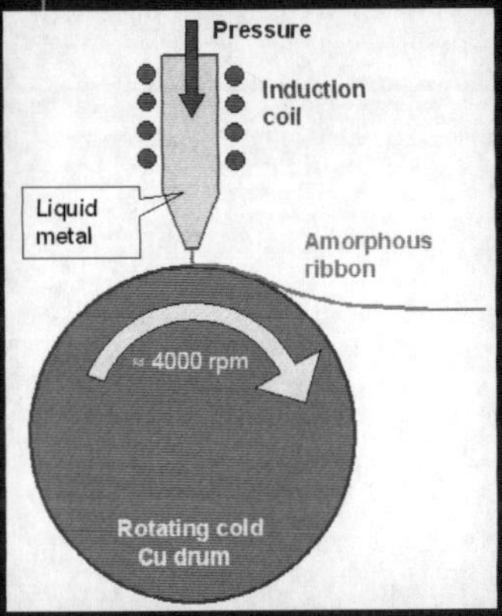

Melt spinning; This strips of 0.03to0.1mm and less than 300mm wide ribbons. Mostly laboratory work.

CVD process- $SiCl_4$ vapor on a 3%silicon sheet and annealing at 1200C.Toxicity issues

Hot dipping: Hot dip Fe-3Si strip in molten AlSi hypereutectic at 800C and diffusion anneal at 1250C. Al reduces induction and magnetostriction

PVD: electron beam magnetron sputtering of silicon vapor on Fe-3%Si strip –porous interfaces are a problem

Spray forming Silicon vapor using argon carrier gas on Fe-3Si strip and annealing at 1050C for 10Hours.

6.5% SiFe

Parameter Optimization

POWDER CHARACTERISTICS

Sample	AD	Hall	Carney	C	S	O	N	d10	d50	d90
6.5 % Silicon Iron- Gas Atomized	4.12	15.7	3.4	.005	.003	.060	.006	15.7	25.7	42.1

Sample	+80	+100	+140	+200	+230	+325	Pan	Tap Den
6.5 % Silicon Iron- Gas Atomized	0	0	T	0.1	T	0.7	99.2	4.35

TEST PARAMETER COMBINATIONS (LAYER THICKNESS 0.04MM)

Test	Energy Input [J/mm^3]	Laser Power [W]	Scanning Speed [mm/s]	Hatch Distance [mm]
1 (316 Std.)	**54.5**	214	982	0.10
2	**89.3**	200	700	0.08
3	**78.6**	220	700	0.10
4	**64.7**	220	850	0.10
5	**55.0**	220	1000	0.10
6	**85.7**	240	700	0.10
7	**70.6**	240	850	0.10
8	**60.0**	240	1000	0.10
9	**92.9**	260	700	0.10
10	**76.5**	260	850	0.10
11	**65.0**	260	1000	0.10
12	**100.0**	280	700	0.10
13	**82.4**	280	850	0.10
14	**70.0**	280	1000	0.10

Test	Energy Input [J/mm^3]	Laser Power [W]	Scanning Speed [mm/s]	Hatch Distance [mm]
15	**68.1**	214	982	0.08
16	**45.4**	214	982	0.12
17	**102.9**	280	850	0.08
18	**60.8**	214	1100	0.08
19	**53.5**	214	1250	0.08
20	**48.6**	214	1100	0.1
21	**62.5**	240	1200	0.08
22	**50.0**	240	1200	0.1
23	**53.4**	250	1300	0.09
24	**111.6**	250	700	0.08
25	**94.0**	220	650	0.09
26	**102.6**	240	650	0.09
27	**111.1**	260	650	0.09
28	**115.7**	250	600	0.09

HOEGANAES CORPORATION

POROSITY VS. ENERGY DENSITY (EXAMPLE)

Printed via laser EOS M290

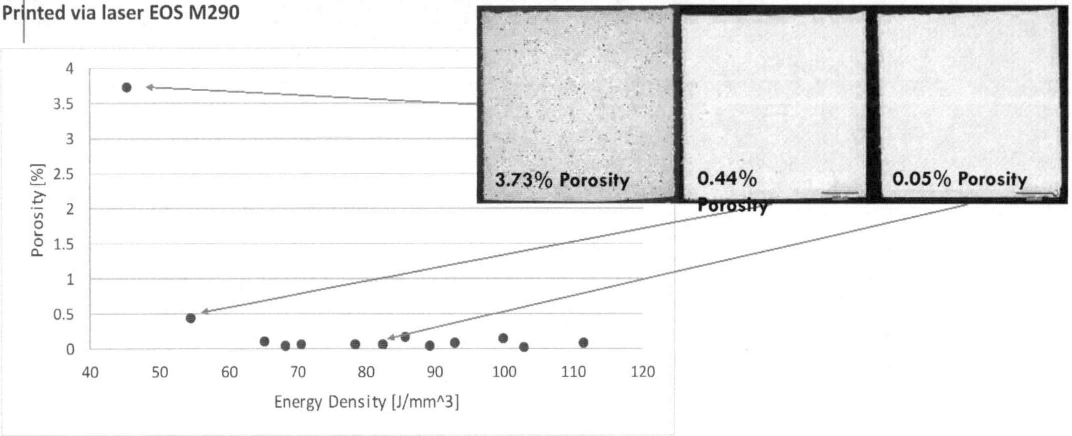

HOEGANAES CORPORATION

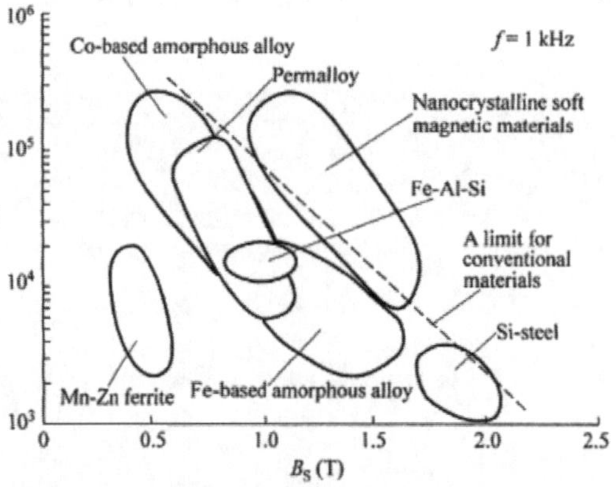

Figure 4. 1 Relationship between relative permeability μ_r and saturation magnetic induction B_S for typical soft magnetic materials.

4 Nanocrystalline Soft Magnetic Materials and Their Applications

Yoshihito Yoshizawa

Yoshihito Yoshizawa "Nano crystalline Soft magnetic Materials and Their applications" Handbook of Advanced Magnetic Materials pp 1452-1486(2006)

PERMANENT MAGNETS

The minimum in energy is when exchange and anistropy energy are equal

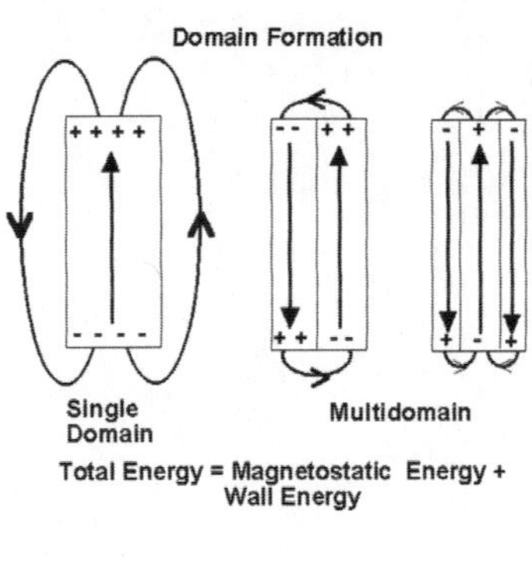

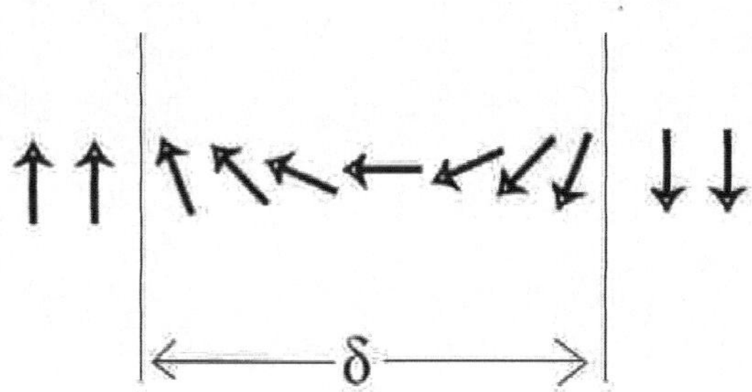

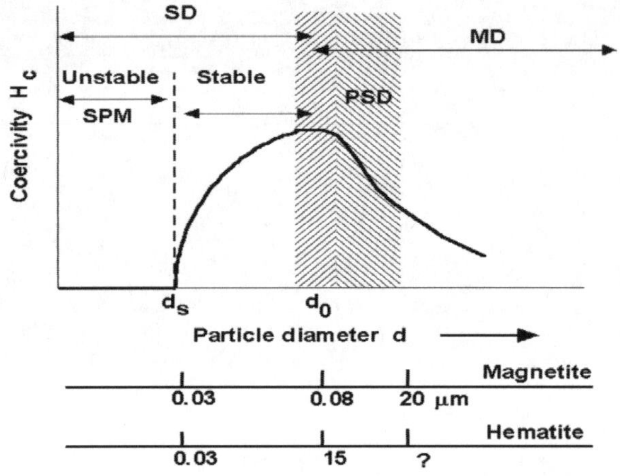

Dunlap 1981

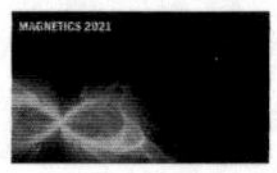

PERMANENT MAGNETIC MATERIALS SINGLE DOMAIN SIZE

Material	Single domain radius(nm)
CoPt	310
MnAl	90
$BaFe_{12}O_{19}$	290
$SmCo_5$	425
Sm_2Co_{17}	250
$Sm_2Fe_{17}N$	190
$Nd_2Fe_{14}B$	107-300
$Pr_2Fe_{14}B$	300

$$r_c = \frac{9\gamma\mu_0}{2 I_s^2}$$

γ is wall energy is product of exchange and Anisotropy

300 nm= 0.3 microns

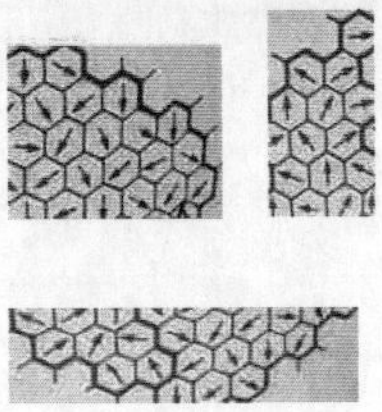

Selective laser melting(SLM) has a cooling rate of 10^5 to 10^7K/sec can Form Fine nano size grain microstructure close to single domain size. Melt Spinning has cooling rate equivalent or better

Melt spun ribbons generally are blended with polymer to make shapes. SLM provides a possibility to form dense bodies with no polymers enhancing the induction achieved.

$(Nd_{0.8}Ce_{0.2})_{2.4}Fe_{12}Co_2B$ Ribbons
Heyun Li, Yang Liang, Xiaohua Tan,* Hui Xu, Pengfei Hu,* and Kezhi Ren

Net Shape 3D Printed NdFeB Permanent Magnet

J. Jaćimović[1*], F. Binda[1], L. G. Herrmann[1], F. Greuter[1], J. Genta[1], M. Calvo[1], T. Tomse[2,3], R. A. Simon[1]

[1]*ABB Corporate Research Center, Baden-Daettwil, Switzerland*
[2]*Department for Nanostructured Materials, Jožef Stefan Institute, 1000 Ljubljana, Slovenia*
[3]*Jožef Stefan International Postgraduate School, 1000 Ljubljana, Slovenia*

Visit toABB

(R-L)
Thomas Cristen, Jacim Jacimovic, Sim, Levina Schaerf

An ABB Synchronous Reluctance motor. Modified from http://quintecgmbh.com/us/news/161-news-abb-synchronour-reluctance-motor.html

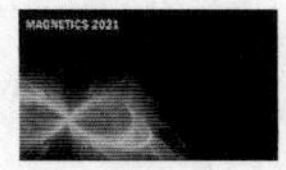

SLM FOR NDFEB MAGNETS

Spherical powders are used to improve flowability to make a dense powder bed. Commercial MQP-S spherical powder of Nd7.5Pr0.7ZR2.6Ti2.5Co2.5Fe75B8.8 from Magnaquench was used .

The powder size distribution had d50 of 38micron with a flow rate of 3 .33g/cm3 and tap density of 4.67 g/cm3 Which is 61.4%of the full density

LP=1700mA.LF=30micron,PD=30micron,ET 110 microsecond ,LT=20micron,HD=100micron

As printed magnets had Br of 0.59T,BHmax of 45kJ/m3(5.65MGOe) and grain size of 0.5 to 1micron scale

J. Jaćimović[1*], F. Binda[1], L. G. Herrmann[1], F. Greuter[1], J. Genta[1], M. Calvo[1], T. Tomse[2,3], R. A. Simon[1]

[1]*ABB Corporate Research Center, Baden-Daettwil, Switzerland*
[2]*Department for Nanostructured Materials, Jožef Stefan Institute, 1000 Ljubljana, Slovenia*
[3]*Jožef Stefan International Postgraduate School, 1000 Ljubljana, Slovenia*

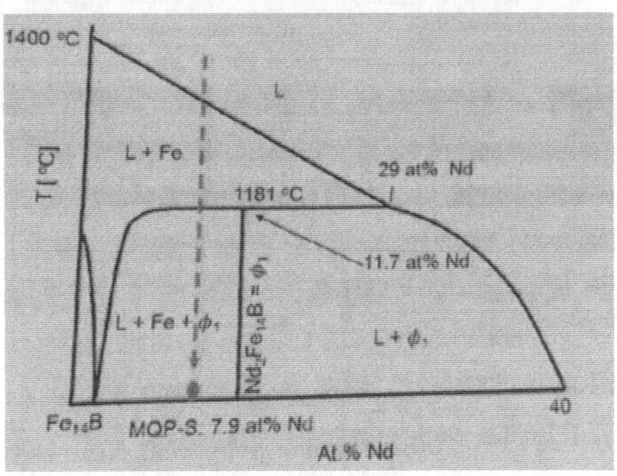

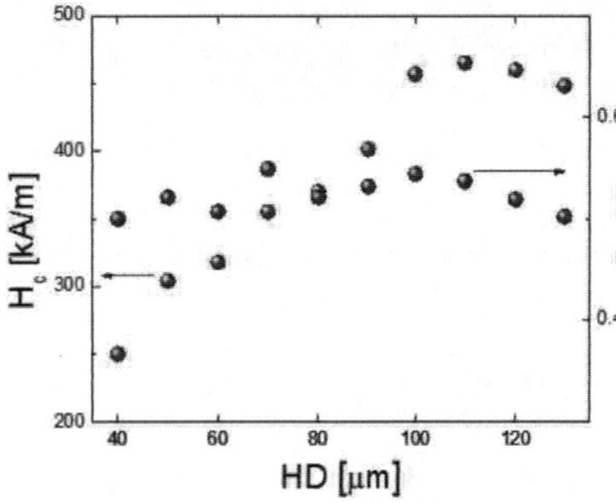

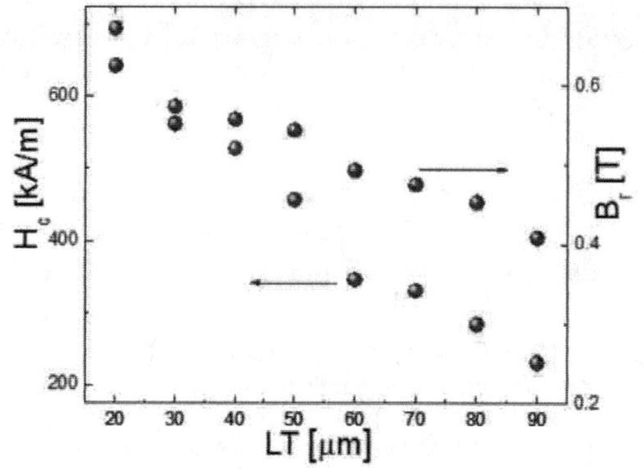

700kA/m =8778 Oe

450kA/m= 5643 Oe

Jacimovic; ABB

Data on $5mm^3$ samples

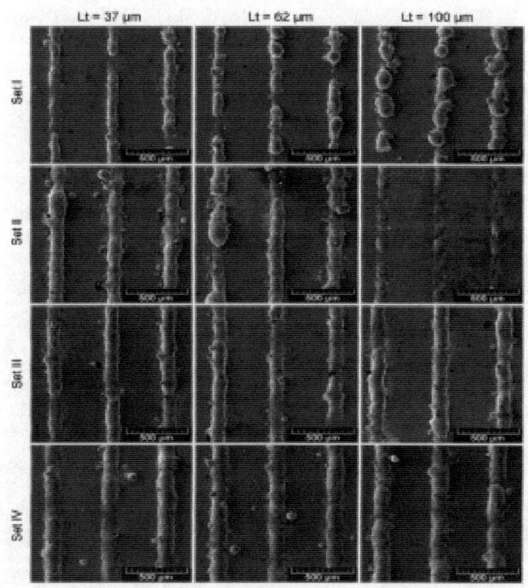

Materials **2020**, *13*, 139; doi:10.3390/ma13010139

Figure 4. Overview of single tracks formed by parameters sets: I, II, III, and IV at various powder layer thickness (Lt).

Feature	Coercivity H_{cJ}	Remanence J_r	Maximum Magnetic Energy Product (BH)max	Relative Density/%	Method
	kA/m	T	kJ/m^3	%	
This study (max. values)	516 [a]	0.563 [b]	35.9 [c]	90.9 [d]	LPBF
Jacimovic et.al	775.3	Max. 0.478	39.78	92	LPBF
Kolb	N/A	Avr. 0.45 Max. 0.514	N/A	86	LPBF
Urban et.al	915	0.587	52.1	N/A	LPBF
[22] without grain boundary infiltration	520	0.436	N/A	65	LPBF
[22] with grain boundary infiltration with Nd50Tb20Cu30	1207	0.390	N/A	N/A	LPBF

Materials **2020**, *13*, 139; doi:10.3390/ma13010139

Influence of Melt-Pool Stability in 3D Printing of NdFeB Magnets on Density and Magnetic Properties

Mateusz Skalon [1], Michael Görtler [2], Benjamin Meier [1,2], Siegfried Arneitz [1,*], Nikolaus Urban [3], Stefan Mitsche [4], Christian Huber [5], Joerg Franke [3] and Christof Sommitsch [1]

Jaimovicet.al
Adv.eng.materials,; 2017, 19, 1700098
Kolbet.alProceedings of the 2016 6th International Electric Drives Production Conference (EDPC), Nuremberg, Germany
Urban, N.; Kühl, A.; Glauche, M.; Franke, J. Additive Manufacturing of Neodymium-Iron-Boron Permanent Magnets. In Proceedings of the 2018 8th International Electric Drives Production Conference (EDPC), Schweinfurt, Germany, 4–5 December 2018; pp. 1–5.

Table 2. A comparison of a typical grain size of NdFeB magnets manufactured with various methods.

Manufacturing Method	Typical Grain Size
Sintering	10–15 µm [11]
Hot pressing and hot deformation	0.3 µm [26]
LPBF	0.5–3 µm (current study)

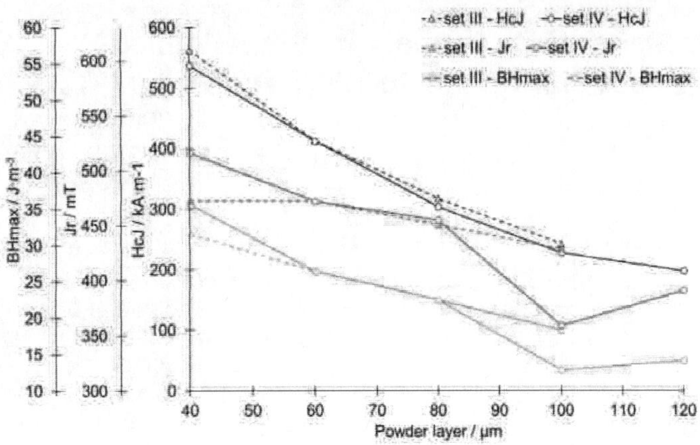

MAGNETICS 2021

BINDER JET PRINTED MAGNETS OF (SMFEN+NDFEB)

Sample	Temperature (K)	H_c (kOe)	B_r (kG)	$(BH)_{max}$ (MGOe)
S1	300	10.8	3.7	2.8
S2	458	11.1	4.7	4.6
S3	468	11.1	6.9	10.2
S4	478	11.0	7.2	11.0

S1=As printed,S2,S3,S4 magnetized at temp at 15KOe field

Kinjal Gandha,Linhg Li,I.C.Niebedim,Brian K.Poat,Vlastimil Kunc, Brian C.Sales, James Bell,M.Prans Paranthaman;"Additve Mnaufacturing of anisotropic hybrid NdFeB-SmFeN nylon composite Bonded magnets;Journal of Magnetism Materials 467,(2018)8-13.

MAGNETICS 2021

SELECTIVE LASER MELTING OF AL-NI-CO PERMANENT MAGNETS

Emma White from Ames laboratory .USA. produced by alnico magnets with a remanence of 0.9T and coercivit' 183kA/m

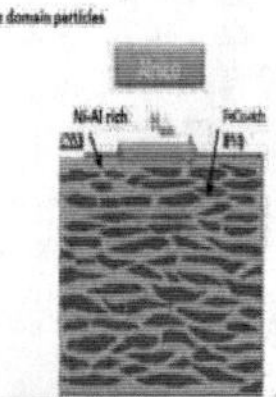

The coercivity is Higher than normal Directional Alnico which generally is about Br of 1.05 to 1.35T and Hc 70kA/m

Directional Alnico 8B(Co35%,Ni15%,Cu 4%,Ti5%,Al7%,Fe34%) has Br of 0.87T and coerci of 127 kA/m

Cast Alnicos process

SELECTIVE LASER MELTING OF AL-NI-CO PERMANENT MAGNETS

Alnico 5-7 (Al8%,Ni14%,Co24%,Cu3%,bal Fe) which has typical values of Br 1.34 T, coercive force of 59kA/m (74 Oe),BH max of 59.7 kJ/m3((7.5MGOe)

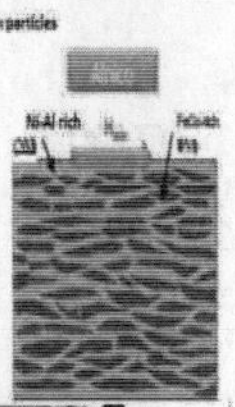

Emma White Data suggest more elongated singl domains increasing the coercive force

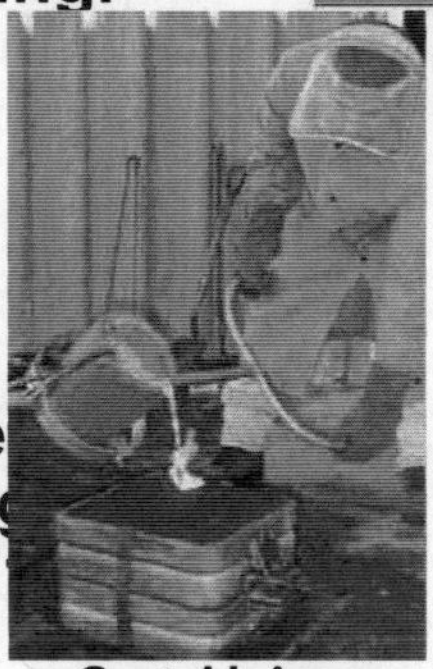

Cast Alnicos process

E. White, A. Kassen, E. Simsek, W. Tang, R. Ott, I. Anderson. Net Shape Processing of Alnico Magne Additive Manufacturing. IEEE Transactions on Ma Volume: PP Issue: 99. DOI: 10.1109/TMAG.2017.27 (2017)

CONCLUSIONS

AM production rate is low compared to the capital investment.

AM is Currently very actively utilized in the aerospace industry. The technology eliminates significant waste of expensive raw material.

In the electrical machines for dc applications AM is very effective as a rapid prototyping tool.

For AC applications Fe-3Si can be made successfully for frequencies below 400Hz,

For medium to higher frequencies significant efforts are underway .

CONCLUSIONS

Binder jet printing offers opportunity but carbon contamination from the binder during debinding is an issue to be resolved for AC applications.

Individual parts in electrical machines have been made successfully ;the complete motor assembly is still a bit faraway.

NdFeB and Rare earth free MnAIC magnets by AM show potential and needs further research

Melt spun rare earth magnets can be replaced by SLM of rare earth alloy powders

METROlab
Magnetic precision has a name

MAGNETICS 2021

Precision Measurement of Time-Varying Fields

Philip Keller
keller@metrolab.com

METROlab
Magnetic precision has a name

Presentation Outline

- Measuring AC magnetic fields
- Visualizing AC magnetic fields
- Analyzing AC magnetic fields
- Mapping AC magnetic fields
- Unsolved problems

Examples:

- Toroidal transformer in base of halogen desk lamp
- 3-Axis fluxgate sensor (Metrolab TFM1186)
- Metrolab EZMag3D software
- Measures earth's field + 50 Hz

Metrolab
Magnetic precision has a name

Measuring AC Magnetic Fields

First adjust data acquisition parameters!

- Acquisition rate
- Averaging
- Block size:
  - Data transfer
  - Data analysis

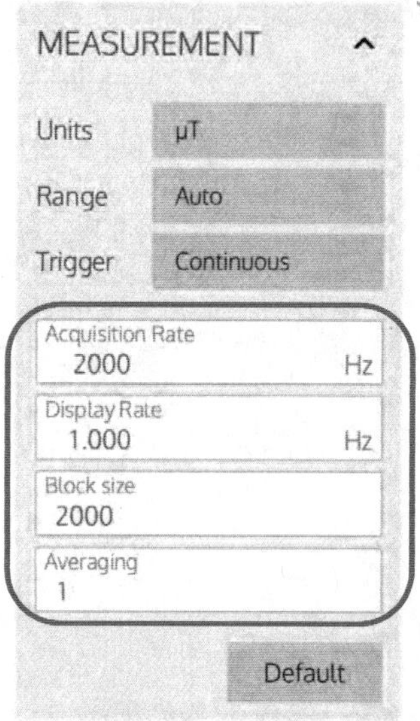

Metrolab
Magnetic precision has a name

Visualizing AC Magnetic Fields

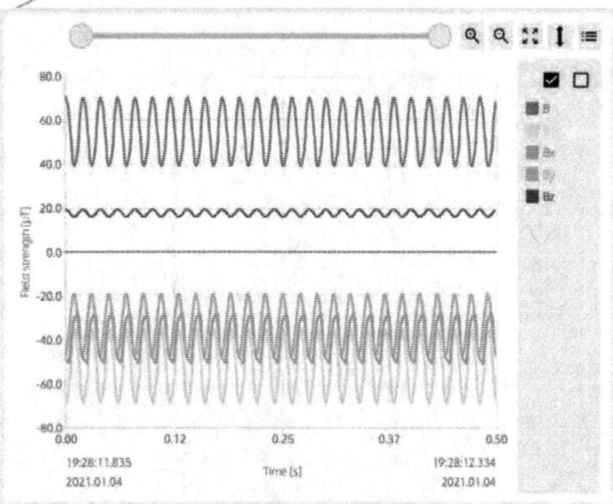

Example: screenshot from EZMag3D

- *Time scale = 0.5 sec*
- *Earth's field with small AC modulation*

Scrolling time plot:

- Select time scale
- Multiple traces:
  - Measured: B_x, B_y, B_z
  - Computed: B, B·B'
- DC offsets
- AC amplitudes
- Wave shape
- "Analytic view"

METROlab
Magnetic precision has a name

Visualizing AC Magnetic Fields

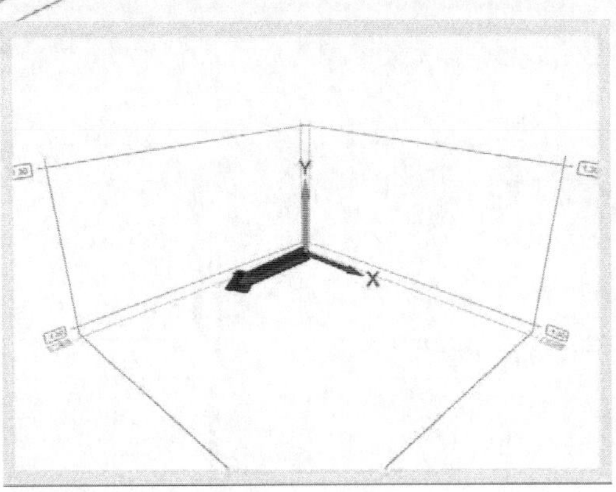

Example: animation in EZMag3D

- *Time warp = 0.01*
- *Earth's field with small AC modulation*

Animated vector plot:

- Select "time warp": real-time, speed up, or slow down
- Combined effect of B_x, B_y, B_z
- Rotate for clarity
- "Intuitive view"

METROlab
Magnetic precision has a name

Analyzing AC Magnetic Fields

- Spectral analysis:
  - Spectral plot
  - List of spectral peaks
- Figures of merit:
  - Root Mean Square amplitude (RMS)
  - Peak-to-peak amplitude (P-P)
  - Standard Deviation (Std. Dev.)
- Computations on measurement blocks

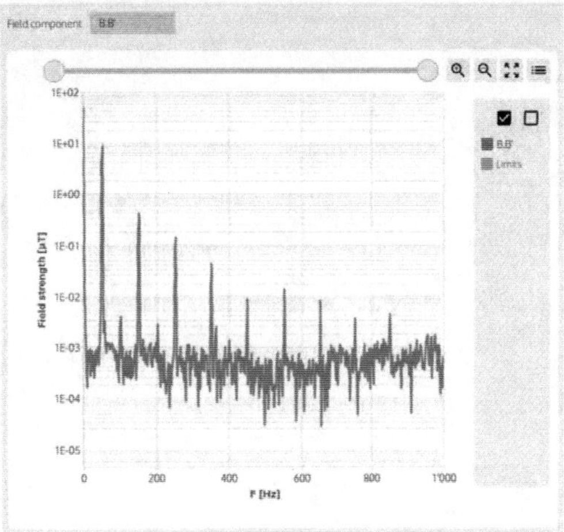

Metrolab
Magnetic precision has a name

Analysis: Spectral Analysis

- What component?
- B_x, B_y or B_z do not capture entire signal
- $B = \sqrt{(B_x^2 + B_y^2 + B_z^2)} \geq 0$: rectification introduces spurious peaks
- B·B': projection of B vector onto direction of field oscillation: as if probe were "auto-aligned" to field oscillation

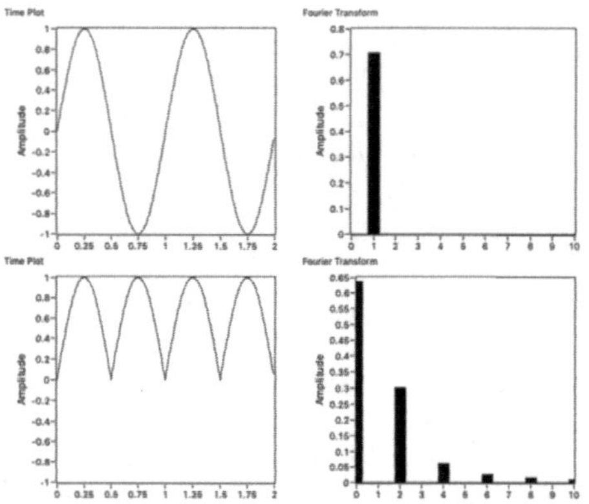

Metrolab
Magnetic precision has a name

Spectral Analysis: B·B' vs. B

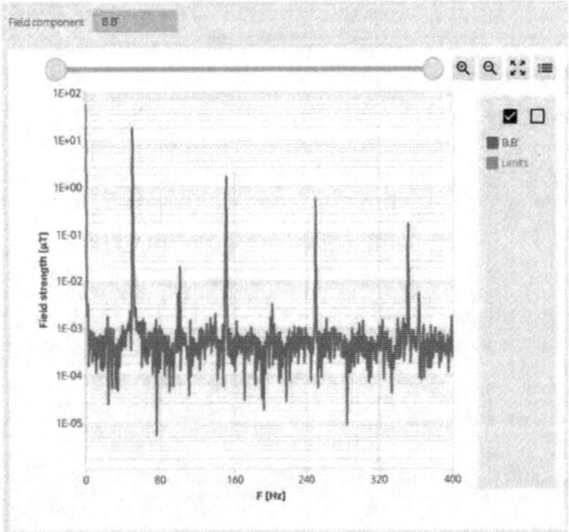

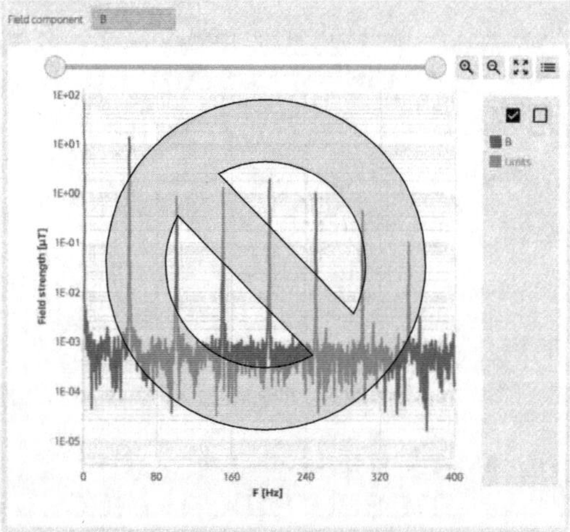

MetroLab
Magnetic precision has a name

Spectral Analysis: Peak Finder

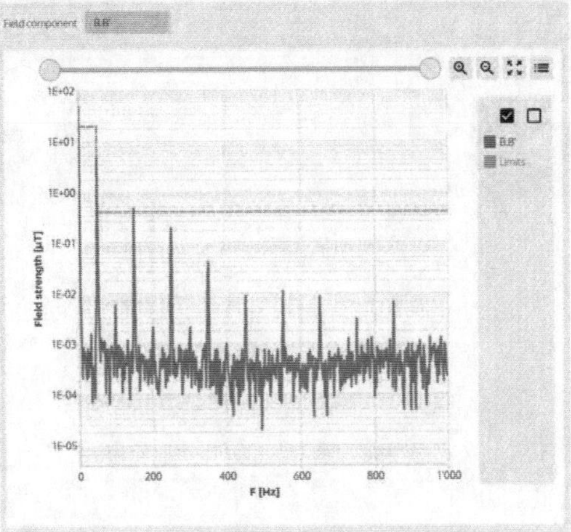

Field component B.B' Threshold 20.00 dB above noise

Peak #	Amplitude [µT]	Frequency [Hz]	Exposure limit [µT]
1	50.53	0.0000	20.00
2	19.11	49.99	20.00
3	0.0234	100.0	0.4000
4	1.876	150.0	0.4000
5	0.3967	249.9	0.4000
6	0.0849	349.9	0.4000
7	0.0183	450.0	0.4000
8	0.0227	549.9	0.4000
9	0.0172	650.0	0.4000
10	0.0138	850.0	0.4000

Example: screenshot of EZMag3D

- *Find spectral peaks in B·B', > 20 dB above noise floor*
- *Exposure limits: 20 µT for 0 … 50 Hz, 0.4 µT for 50 … 1000 Hz*

MetroLab
Magnetic precision has a name

Analysis: Root-Mean Square (RMS)

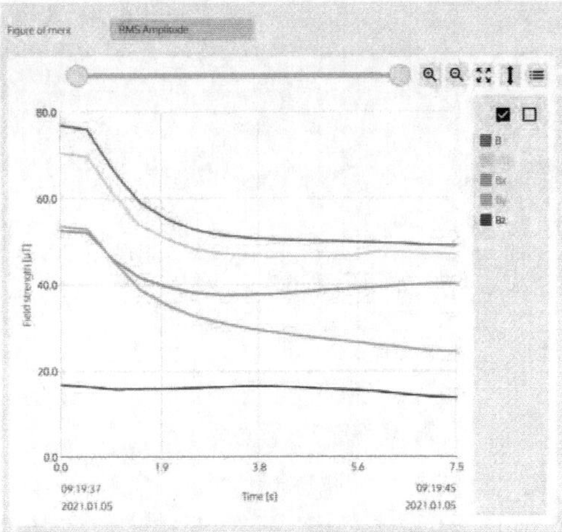

RMS includes DC term (earth's field) = limit as distance increases

METROlab
Magnetic precision has a name

Analysis: Peak-to-Peak

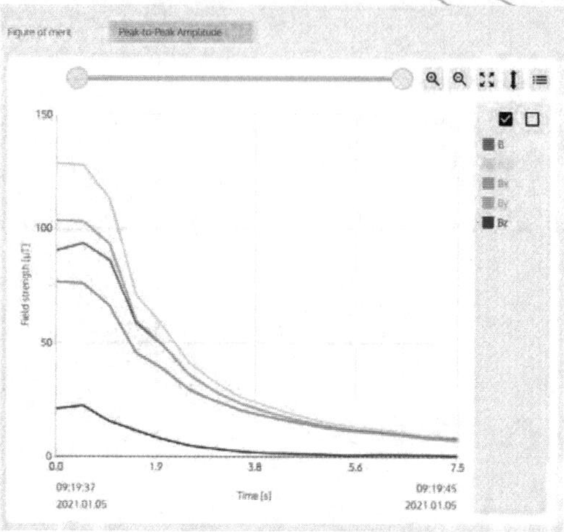

Peak-to-Peak amplitude provides an estimate of the AC component only. Tends to zero as distance increases.

METROlab
Magnetic precision has a name

Analysis: Standard Deviation

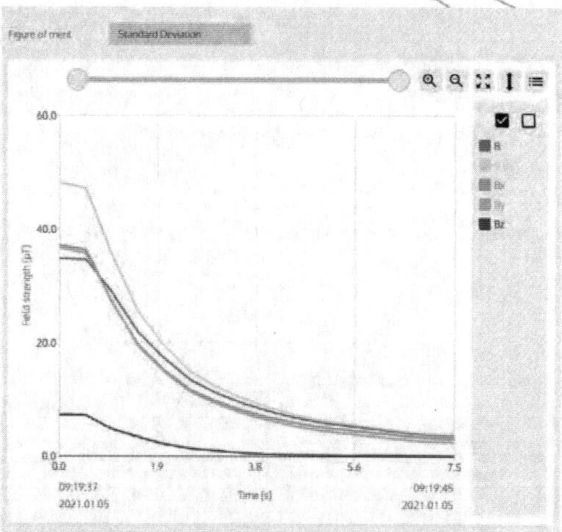

Standard Deviation is like Peak-to-Peak, but is more stable (smoothing effect). Usually, StdDev ~ 1/3 P-P.

METROlab
Magnetic precision has a name

Analysis: By Frequency Band

Figure of merit: Standard Deviation

Freq. Band [Hz]	B [µT]	B.B' [µT]	Bx [µT]	By [µT]	Bz [µT]	Eposure Limit [µT]
0-50	0.0140	0.0157	0.0163	0.0166	0.0164	20.00
50-1000	3.660	3.800	3.196	2.547	0.1070	0.4000

Example: screenshot from EZMag3D

- *Compute Standard Deviation of all components by frequency band*
- *Measured signals (B_x, B_y, B_z) as well as computed signals (B, B·B')*
- *Exposure limits: 20 µT for 0 … 50 Hz, 0.4 µT for 50 … 1000 Hz*

METROlab
Magnetic precision has a name

Mapping AC Magnetic Fields

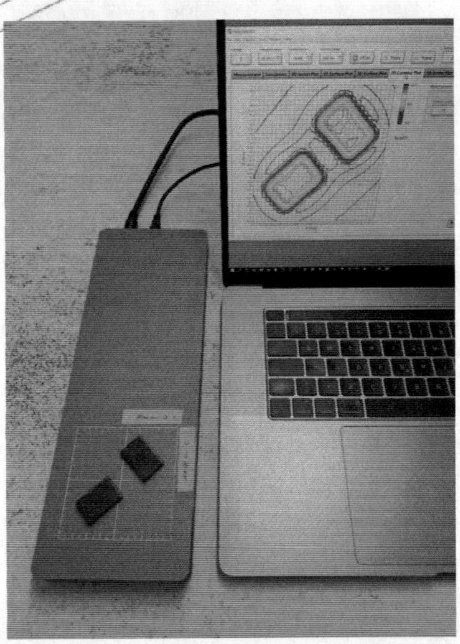

Fraunhofer HallinSight™ array of 3-axis Hall sensors:

- Array: 32 x 32 @ 2.5 mm
- Area: 80 x 80 mm
- Time to map: 40 ms
- Ranges: 0.1, 0.4, 0.8, 2 T
- Resolution: 4 µT
- Accuracy: 0.1%
- Position error: < 50 µm

METROlab
Magnetic precision has a name

Mapping an AC Coil

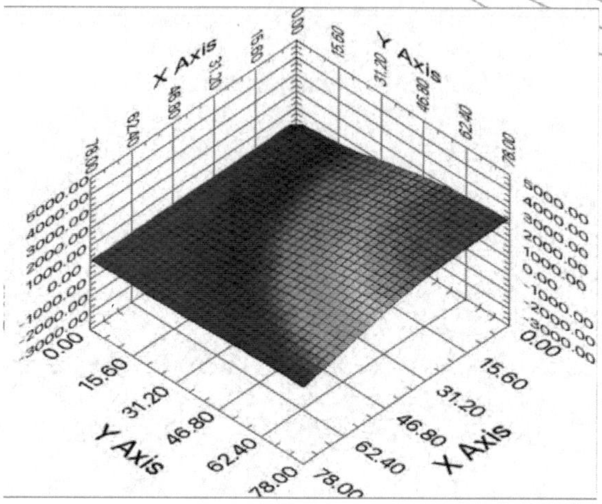

Animation: B_z; real-time; move coil at t = 13 s

METROlab
Magnetic precision has a name

Solved: Definition of RMS(B)

e.g., Directive 2013/35/EU specifies exposure limits as RMS values – but ***B*** is vector; how do we calculate RMS(***B***)??

By analogy of AC power:

$$P = I^2 R \Longrightarrow P_{av} = I_{RMS}^2 R$$

Energy density of magnetic field:

$$u_B = \frac{B^2}{2\mu} \Longrightarrow (u_B)_{av} = \frac{B_{RMS}^2}{2\mu}$$

Then, logically, $(u_B)_{av} = \left(u_{B_x}\right)_{av} + \left(u_{B_y}\right)_{av} + \left(u_{B_z}\right)_{av}$, so:

$$RMS(\boldsymbol{B}) \equiv B_{RMS} = \sqrt{(B_{x\,RMS})^2 + \left(B_{y\,RMS}\right)^2 + (B_{z\,RMS})^2} =$$

$$\sqrt{(RMS(B_x))^2 + \left(RMS(B_y)\right)^2 + (RMS(B_z))^2} =$$

$$\sqrt{\frac{1}{n}\sum_1^n B_{x\,i}^2 + \frac{1}{n}\sum_1^n B_{y\,i}^2 + \frac{1}{n}\sum_1^n B_{z\,i}^2} =$$

$$\sqrt{\frac{1}{n}\sum_1^n \left(B_{x\,i}^2 + B_{y\,i}^2 + B_{z\,i}^2\right)} = \sqrt{\frac{1}{n}\sum_1^n B_i^2} = RMS(B)$$

METROlab
Magnetic precision has a name

Unsolved: Generalized Spectral Analysis

- $\mathcal{F}$(B·B') is limited to linear oscillations
- Far field: usually valid
- But not always valid!!
- Generalized spatiotemporal analysis techniques: complicated, non-intuitive

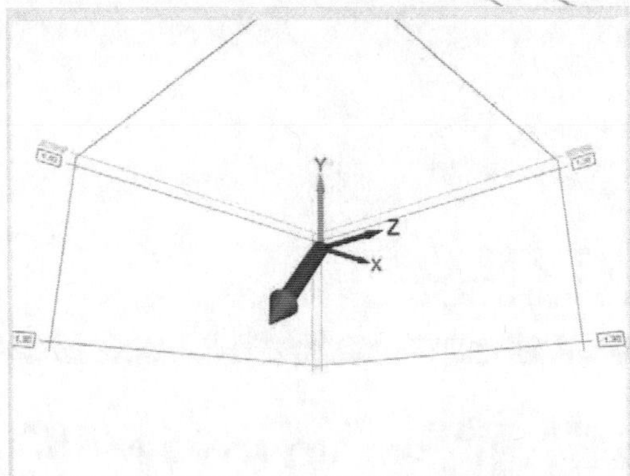

Example: animation in EZMag3D
- *Near field of toroid transformer*
- *Time warp = 0.01*
- *Earth's field with AC modulation*

METROlab
Magnetic precision has a name

Unsolved: AC Calibration

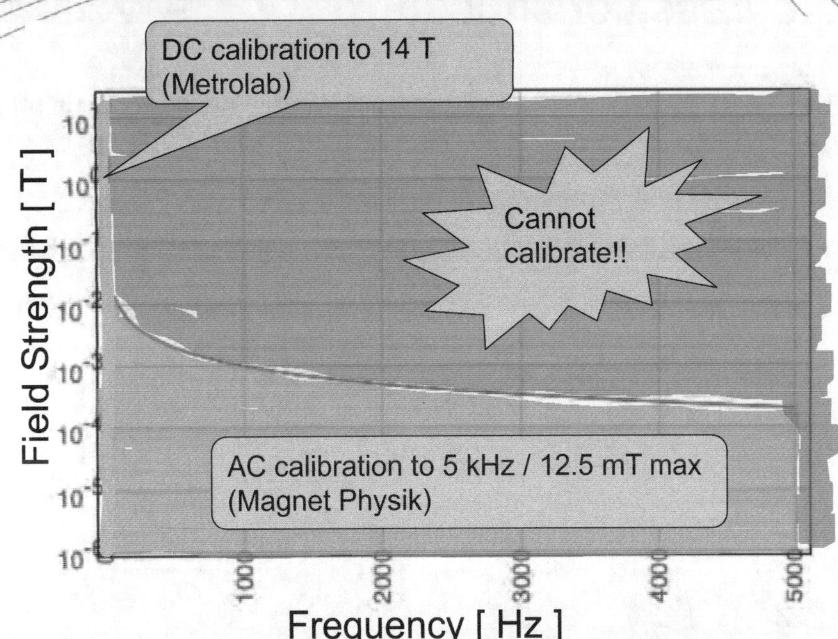

Metrolab
Magnetic precision has a name

AC Magnetic Fields: Conclusion

- Have effective techniques for:
  - Measuring
  - Visualizing
  - Analyzing
  - Mapping
- Challenges remain
- But progress continues!

Overmolding

From magnet to component

MAGNETICS 2021
Martti Paju

Outline

- Assembly of Magnets
- What is Insert molding & Overmolding
  - Process
  - machines
  - tooling
- Benefits
- How to proceed

MAGNETICS 2021
Martti Paju

HANGZHOU PERMANENT MAGNET GROUP

Assembly of magnets

Assembly of metallic magnet

- Glueing can be problematic
- Mechanical assembly as well
- Quality difficult to control

MAGNETICS 2021
Martti Paju

HANGZHOU PERMANENT MAGNET GROUP

Assembly of magnets

Magnets behaving like magnets...

MAGNETICS 2021
Martti Paju

HANGZHOU PERMANENT MAGNET GROUP

Overmolding or Insert Molding

- Insert molding => metallic insert
- Overmolding => plastic insert
- Basically very simple
- Well established technology
- Widely used in injection molding technology
- Not uncommon in magnet technology
- Flexible and scalable
  - Automatisation for large volumes
  - Manual for small or medium size
- Demanding for the mold
- Good process control needed

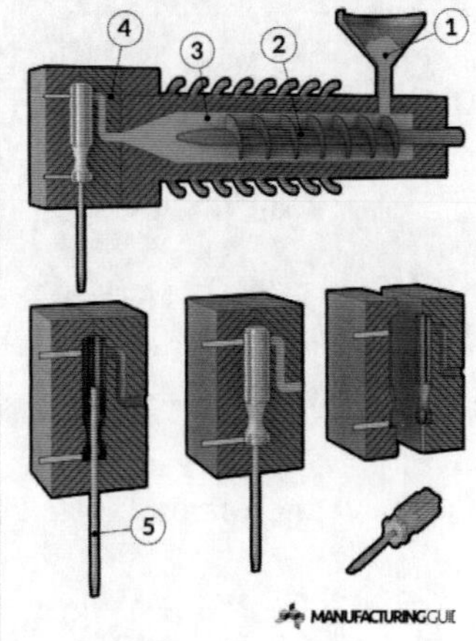

MAGNETICS 2021
Martti Paju

HANGZHOU PERMANENT MAGNET GROUP

Overmolding or Insert Molding

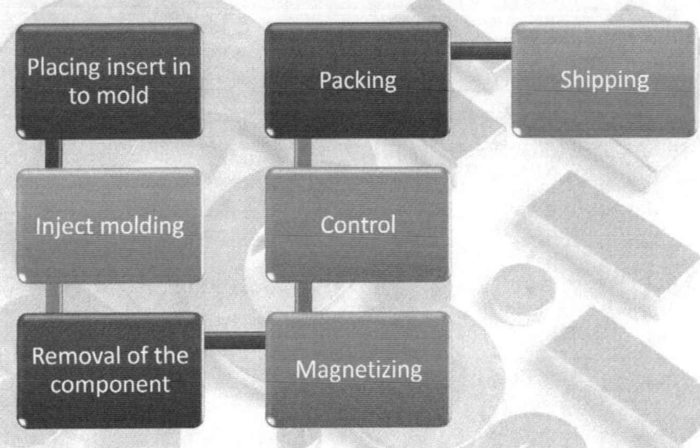

MAGNETICS 2021
Martti Paju

HANGZHOU PERMANENT MAGNET GROUP

Overmolding or Insert molding

For large series

- Fully automatic production cell
- Automatic insert placing
- Automatic removal & control

For small and medium size series

- 2 or 3 station vertical machine
- Manual insert placing
- Manual removal

Overmolding or Insert molding

Molding tool

- Basically normal injection molding tool
- System to keep magnet firmly in place during molding
- Non-magnetic, if the magnets are magnetized
- Tigth tolerances to avoid burrs or flash

Overmolding or Insert molding

Benefits

- Finished component, easy to assemble
- Reduced assembly and labor costs
- Increased reliability
- Increased design flexibility

- Easy magnetic design (no metals)
- Possibly reduced magnet size
- Magnet well protected

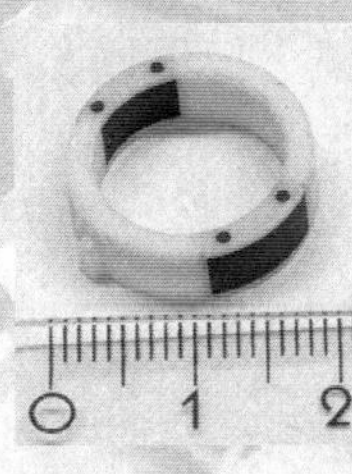

Overmolding or Insert molding

- Best suitable for relatively small components
- Suitable for all magnetic materials, even bonded magnets
- Basically, any plastic material can be used
- Low or no water adsorption polymer best for NdFeB and SmCo
- Freedom of design
- Thermal load by plastic material => magnet grade selection
- Mechanical details may be needed to keep magnet on place

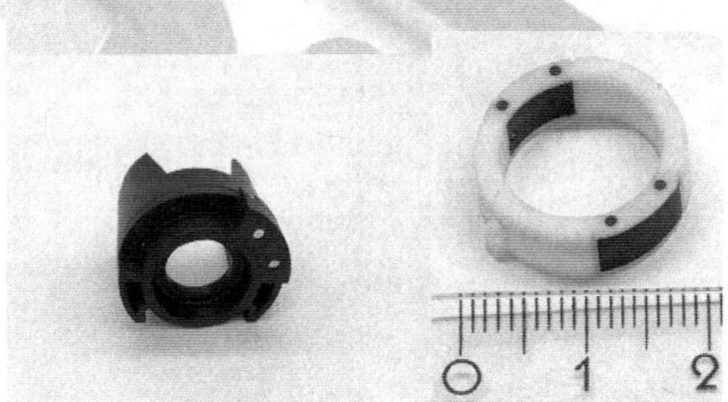

Benefits of overmolding

- Two or more parts can be overmolded at the same time
- Magnet and magnetically active parts
- Magnet and mechanical parts, bearing, bushing etc.

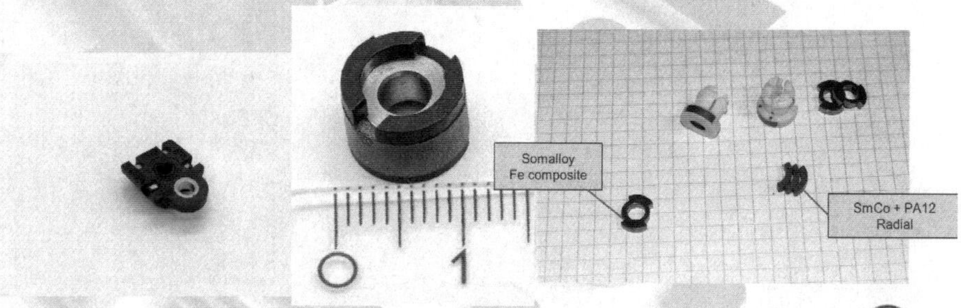

MAGNETICS 2021
Martti Paju

HANGZHOU PERMANENT MAGNET GROUP

Benefits of overmolding

- Functional details easily integrated
- Gears, impeller, lever etc.

MAGNETICS 2021
Martti Paju

HANGZHOU PERMANENT MAGNET GROUP

Benefits of overmolding

- Functionality also by material selection, example swimmer
- Huge selection of plastic materials available

MAGNETICS 2021
Martti Paju

HANGZHOU PERMANENT MAGNET GROUP

Hermetic sealing

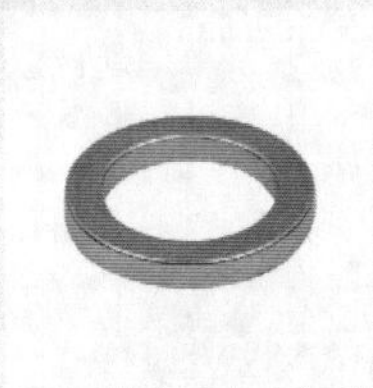

Magnet

After first molding

Final part

- Magnet need to be supported in the mold
- Therefore magnet will be exposed in some places
- Second molding is needed to hermetically seal the magnet

MAGNETICS 2021
Martti Paju

HANGZHOU PERMANENT MAGNET GROUP

RECYCLING

- Overmolded components easy to recycle
- No glue, no other metals
- Plastic and magnet material easy to separate
- Both can be recycled
- Important: Design for recycling

MAGNETICS 2021
Martti Paju

HANGZHOU PERMANENT MAGNET GROUP

IMPORTANT STEPS

- **Same design rules**

- Maintain uniform wall thickness
- No too thin or thick walls
- Avoid sharp corners
- Follow rib design
- Apply a draft

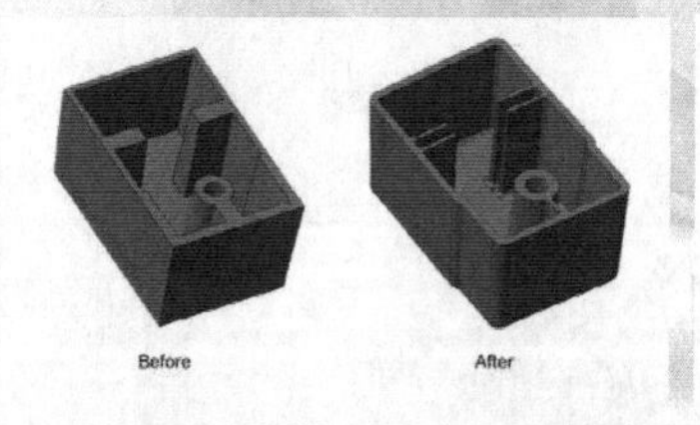

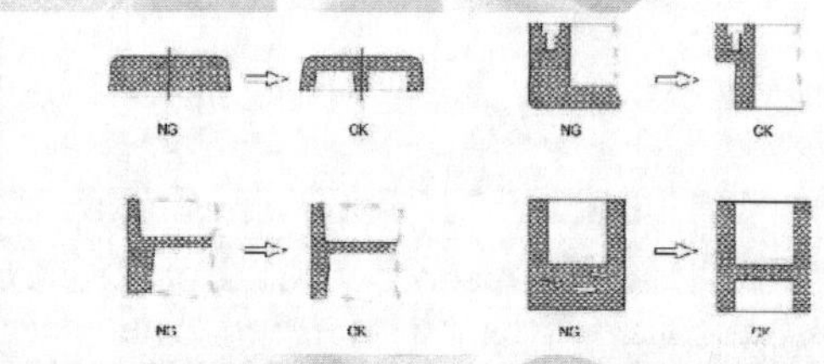

MAGNETICS 2021
Martti Paju

HANGZHOU PERMANENT MAGNET GROUP

IMPORTANT STEPS

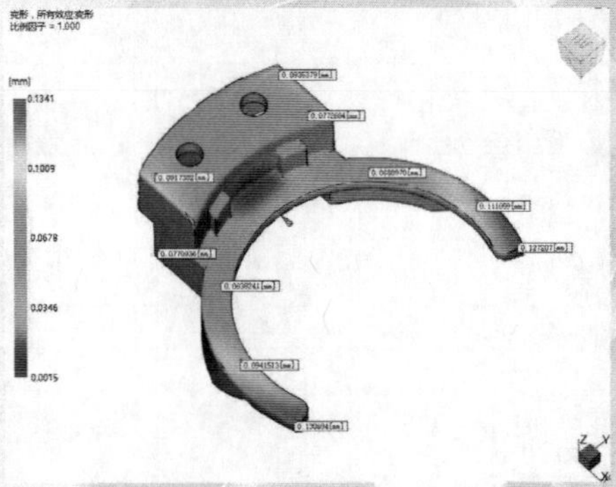

Mold-flow Analysis
- Highly recommended,
- To get right
  - Gate position
  - Wall thickness
  - Balanced filling
  - Share rate & pressure

- to avoid
  - Distortion
  - Shrink marks
  - Weld lines
  - Air traps

MAGNETICS 2021
Martti Paju

HANGZHOU PERMANENT MAGNET GROUP

MOULDING ANALYSIS

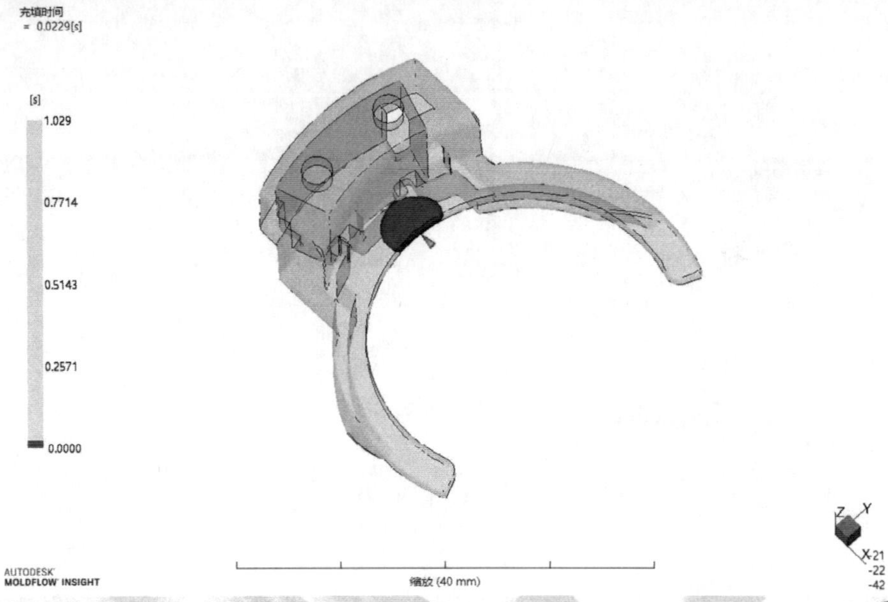

MAGNETICS 2021
Martti Paju

HANGZHOU PERMANENT MAGNET GROUP

Moulding analysis

内容 content	标准允许值 standard	分析结果 analysis results	永磁结论 HZPMG conclusion	客户结论 Customer conclusion
充填 filling	平衡 balanced	平衡	**OK**	
流动前沿温度 (℃) Temperature at flow front	190~230℃	206.3~210.7 ℃	**OK**	
最大注射压力 (MPa) Max injection pressure	<80MPa	11.68 MPa	**OK**	
最大锁模力 (T) max clamp force	待定 TBD	5 T	**OK**	
最大剪切速率 (1/s) Max Shear rate	40000 1/s	16161 1/s	**OK**	
熔接痕 weld line	有 Yes	yes	**Can be optimized**	
困气 air trap	无No	无 No	**OK**	
充填时间 filling time		1S	**OK**	
保压时间 holding time		6s	**OK**	
冷却时间 cooling time		8S	**OK**	
变形 Deflection all effects(mm)		Top thin wall deformation about 0.024mm	需客户确认	

Moulding analysis

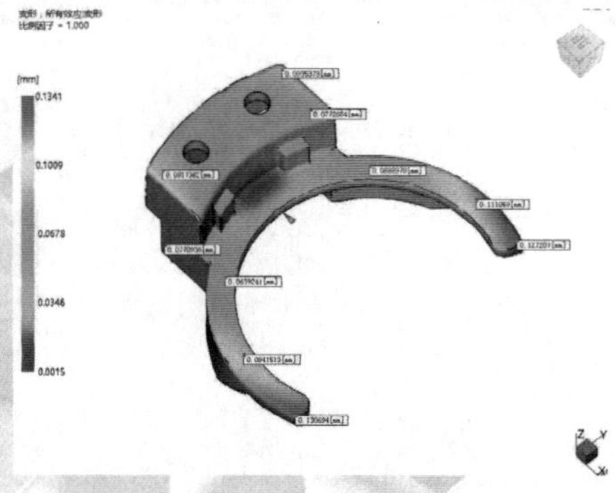

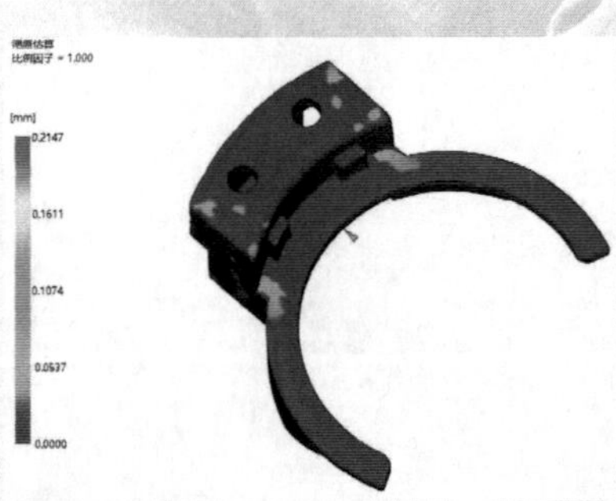

Possible problems with:

- Weld lines
- Shrinkage
- Deflection

MAGNETICS 2021
Martti Paju
HANGZHOU PERMANENT MAGNET GROUP

Magnetic analysis

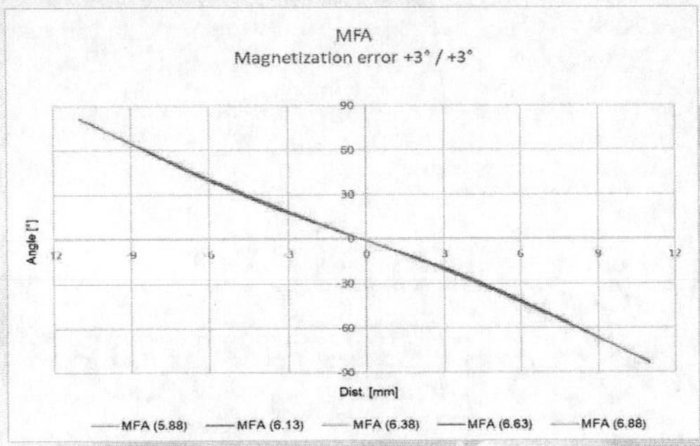

Magnetic analysis

- Additional to normal
  - Angular error due to possible shape distortion
  - Angular error due to magnet movement

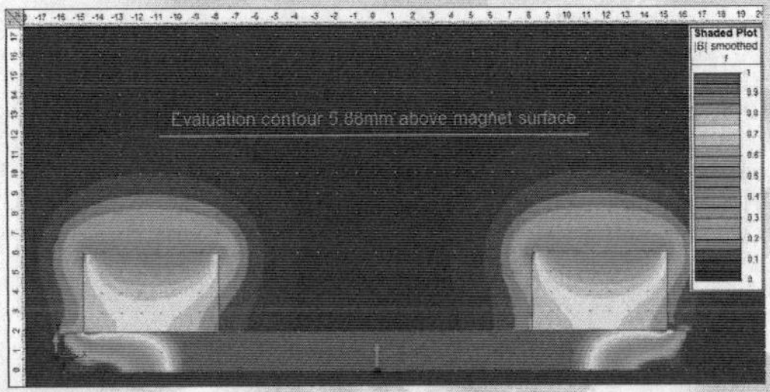

Thank you very much for your attention!

Estimating Complete Hysteresis Loops using Excel

Stan Trout

January 20, 2021

Spontaneous Materials

6 feet

Spontaneous Materials

Outline

- Background
- The Problem
- Prior Art
- The Model
- Comparisons to Reality
- Conclusions

Spontaneous Materials

Background

Spontaneous Materials

Background

Stops along the way
(Where I learned about magnets)

- Univ. of Pennsylvania
- Recoma
- Crucible
- Hitachi
- Magnequench
- Molycorp
- Consulting

20 Years

Spontaneous Materials

What Spontaneous Materials Does

- Technical Training
- Material Selection
- Industry Reports
- Process Development
- Troubleshooting
- Magnetic Design Review
- Due Diligence
- Expert Witness
- Advice to Investors
- Market Studies

Third edition

Contact me if you have any suggestions for the update

Spontaneous Materials

Permanent Magnets 2021-2031

- Fully updated, from beginning to end
- A market study written by experts, not by AI
- Scheduled for mid-2021
- Authors
  - Walt Benecki
  - Steve Constantinides
  - John Ormerod
  - Stan Trout

Spontaneous Materials

The Problem

Spontaneous Materials

Textbook

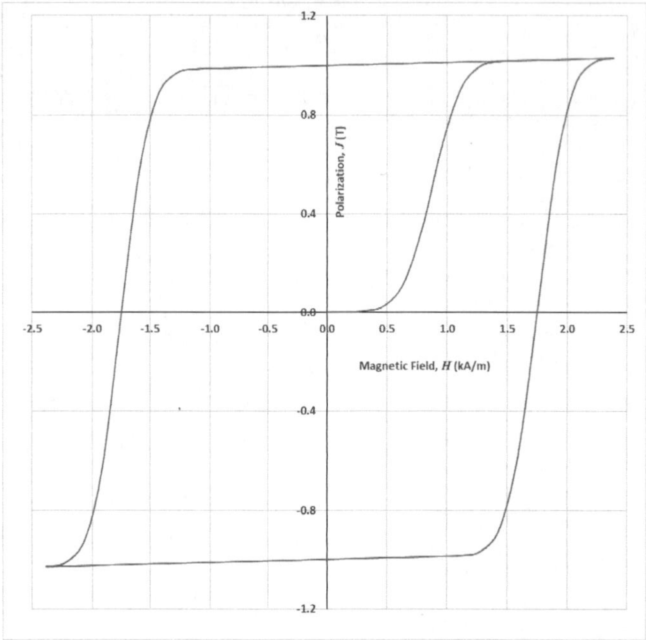

Spontaneous Materials

Datasheet

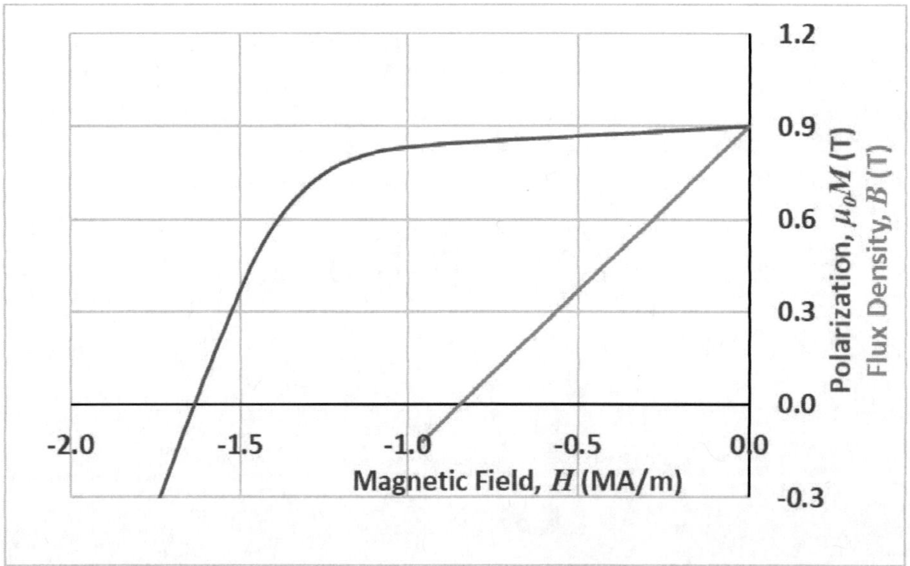

Spontaneous Materials

Production

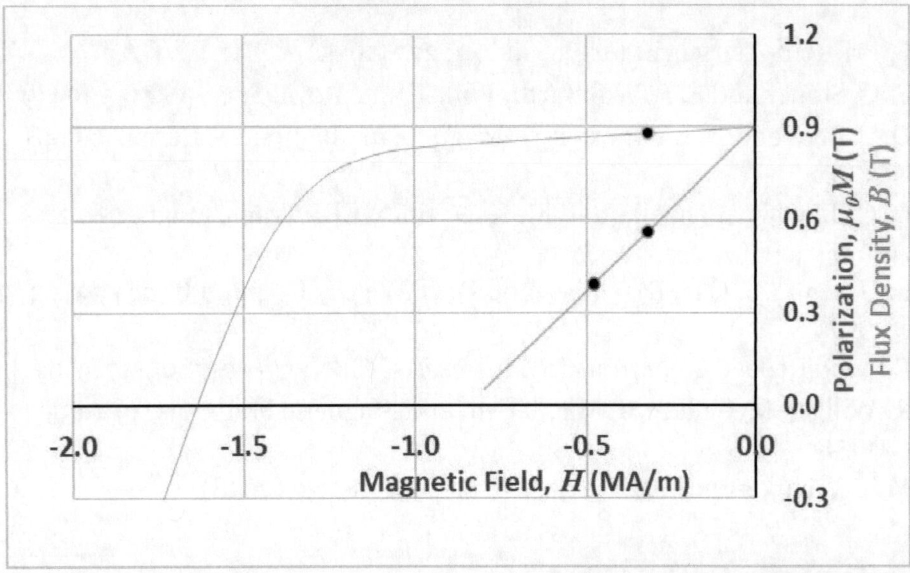

Spontaneous Materials

Prior Art

Spontaneous Materials

Prior Art

- F. Priesach, Zeitschrift für Physik **94**, 277 (1935)
- E. C. Stoner and E. P. Wohlfarth, Phil. Trans. Royal Soc., **A240** 599 (1948)
- D. C. Jiles and D. L. Atherton, J. Magnetism Magnetic Materials **61**, 48 (1986)
- D. C. Jiles, J. B. Thoelke and M. K. Devine, IEEE Trans. Magn. **28**, 27 (1992)
- L. H. Lewis, J. Gao, D. C. Jiles and D. O. Welch, J. Appl. Physics **79**, 6470 (1996)
- Edward Della Torre, *Magnetic Hysteresis*, ISBN: 978-0-7803-4719-9 (1998)
- R. William McCallum, J. Magnetism and Magnetic Materials **292** 135 (2005)
- M. F. Campos and J. A. De Castro, SMMM '18 737 (2018)

Spontaneous Materials

Prior Art

- The prior art tries to derive the hysteresis loop from first principles and *intrinsic* properties
  - M_s, H_A (anisotropy field)
  - Elegant
- This paper tries to estimate full hysteresis loops from four *extrinsic* (process dependent) properties
  - B_r, H_k, H_{cJ} (H_{ci}) and μ_r
  - Pragmatic

Spontaneous Materials

The Model

Spontaneous Materials

Assumptions

- Grains are non-interacting
- All grains are the same size
- Each grain has a unique field value at which it reverses its magnetization, H_{flip}
- H_{flip} values have a Gaussian (normal) distribution
  - Average = H_{cJ}, where 50% of the grains have flipped
  - H_k is where 5% of the grains have flipped
  - Standard deviation, $\sigma = \dfrac{H_{cJ} - H_k}{1.645}$
- The slope of the $\mu_0 M$ vs. H curve near B_r is related to alignment
  - Slope = 0 for perfectly aligned samples
  - Slope → 1 for isotropic material
  - The slope is (μ_r-1)

Spontaneous Materials

Reasonable Assumptions?

- Weak interactions between grains Most significant
- Grains are not the same size or shape
  - Likely not critical
- H_{flip} likely is unique and obeys a normal distribution
- The slope of the $\mu_0 M$ vs. H curve near B_r may not be a constant everywhere
  - May be a weak function of H

Spontaneous Materials

Normal Distribution

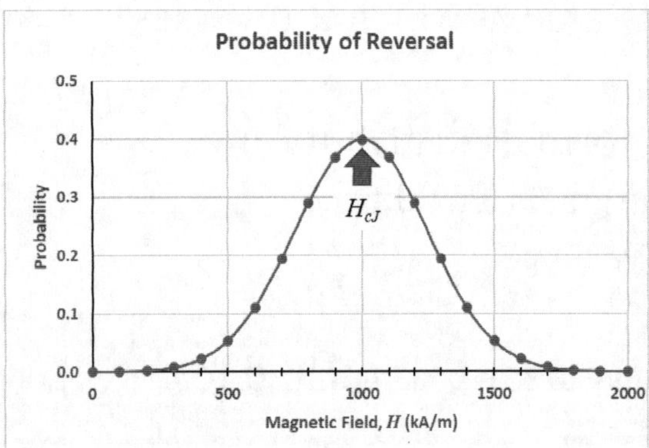

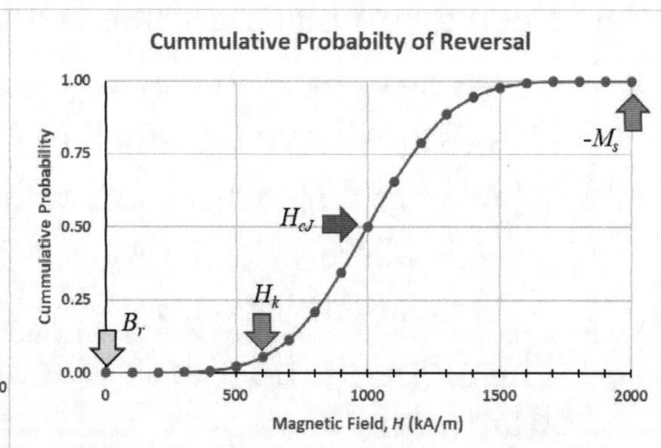

Spontaneous Materials

Ideal Behavior

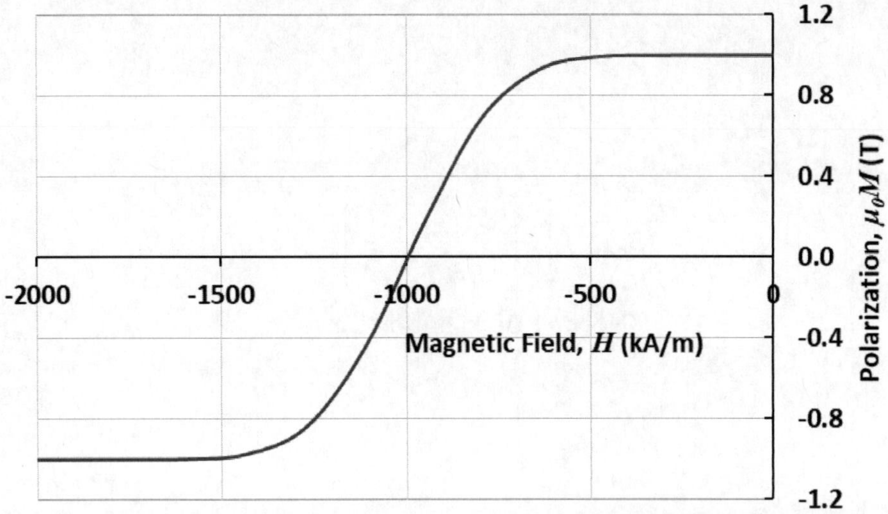

Spontaneous Materials

Complete Loop

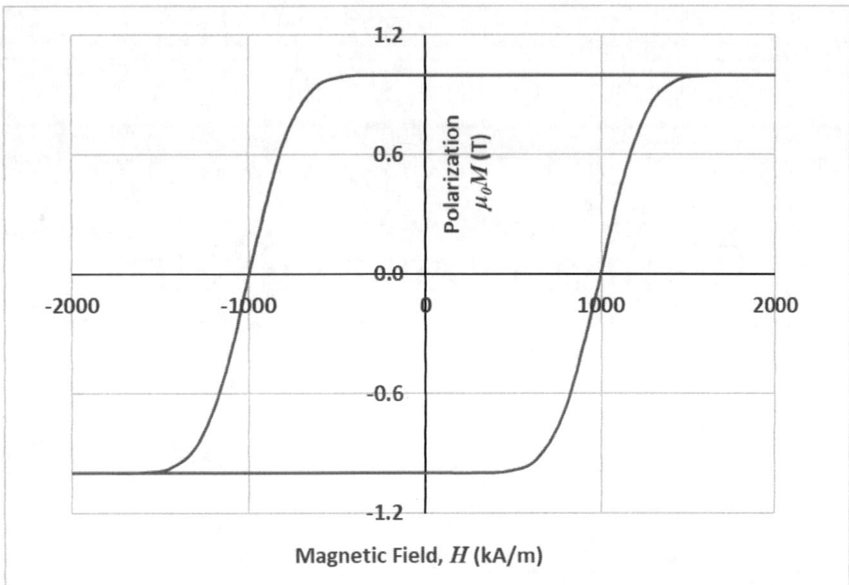

Spontaneous Materials

Equation

Alignment Term

$$J = (B_r)erf\left[\frac{1.645(H_{cJ} - H)}{\sqrt{2}(H_{cJ} - H_k)}\right] + \mu_0 H(\mu_r - 1)$$

Normal Distribution Term

Spontaneous Materials

Comparisons to Reality

Spontaneous Materials

Sintered NdFeB

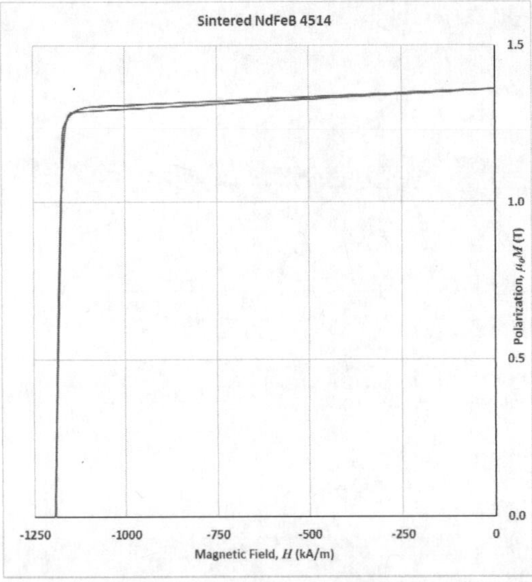

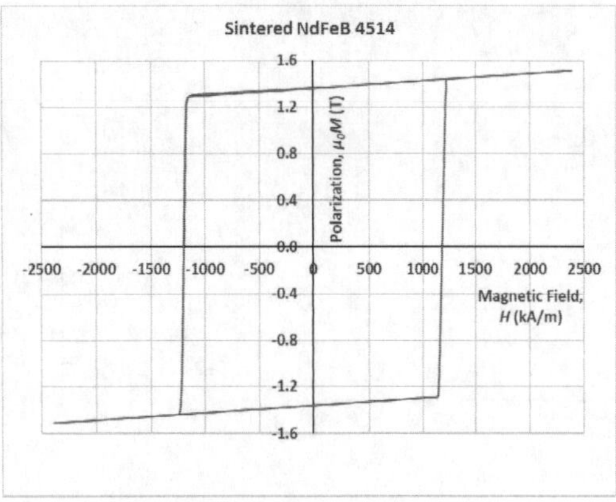

Correlation =0.99

Blue=Calculated
Red=Measured

Spontaneous Materials

Bonded NdFeB

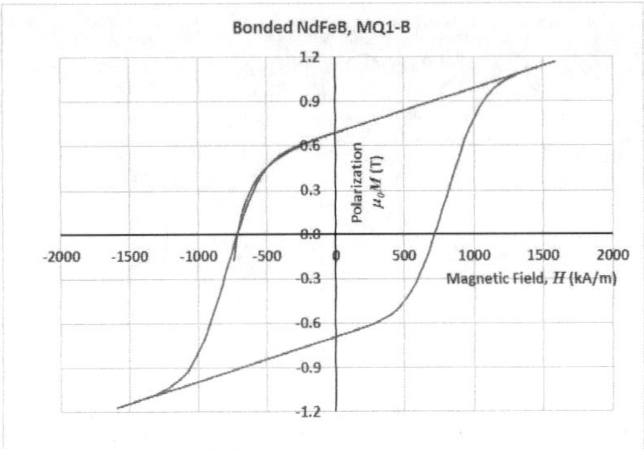

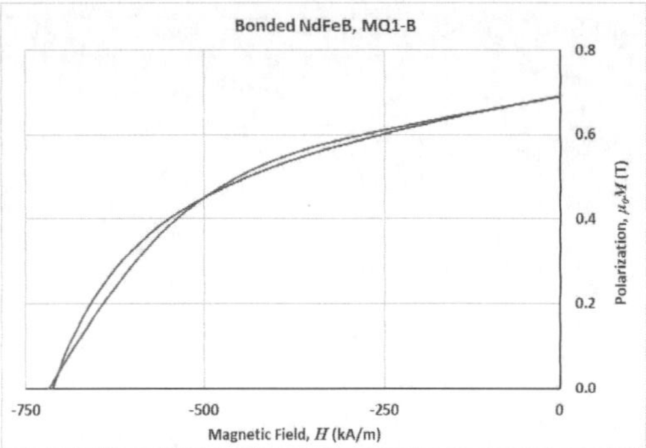

Blue=Calculated
Red=Measured

Correlation =0.99

Spontaneous Materials

SmCo 2-17

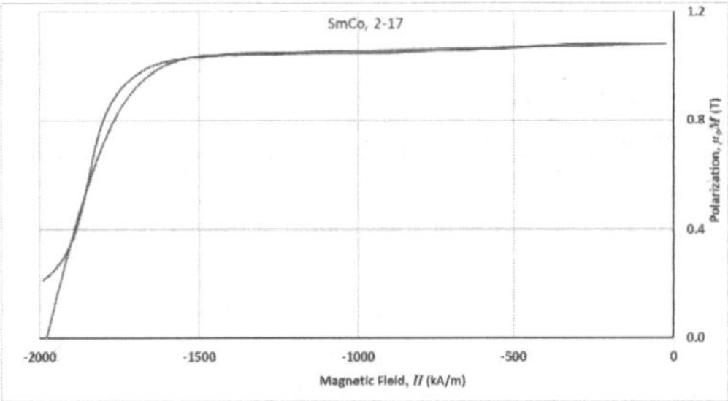

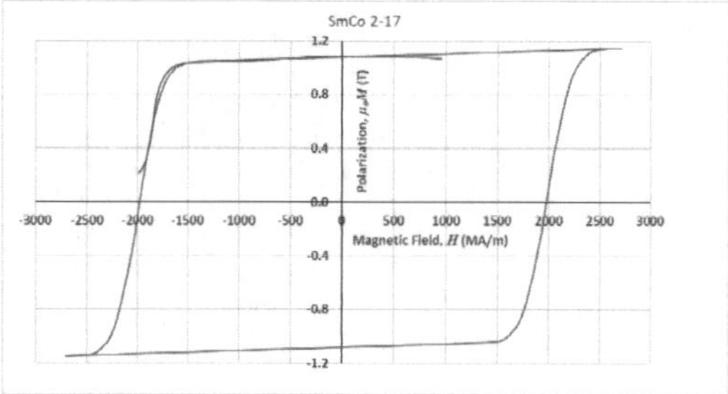

Blue=Calculated
Red=Measured

Correlation =0.99

Spontaneous Materials

Ferrite

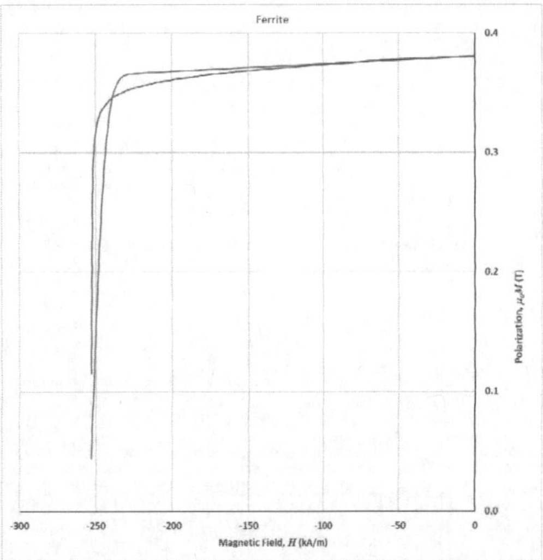

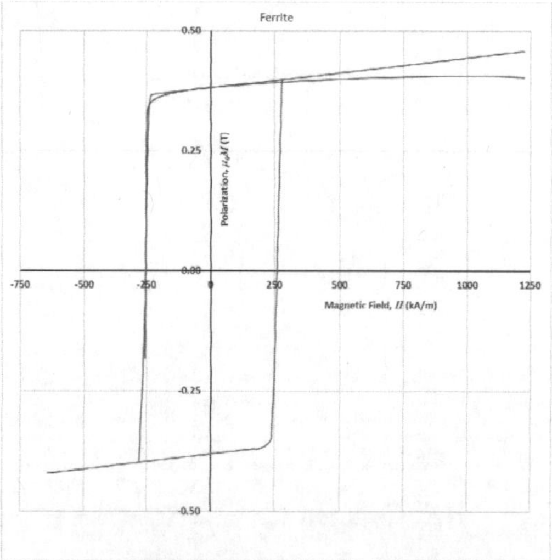

Blue=Calculated
Red=Measured

Correlation =0.89

Spontaneous Materials

Conclusions

Conclusions

- The model shows good fit to a few real curves when properly adjusted, correlation > 89%
  - Rare earth magnets correlate better than ferrites
- This technique could be adapted to improve the quality of measurements
  - Large production quantities
  - H_k and H_{cJ} are convenient, not essential
  - Three points on the curve
- Suitable for artwork
- Spreadsheet template is available strout@ieee.org

Spontaneous Materials

Thanks for the data

- Magnequench
- Vacuumschmelze
- Magnet Physics

Spontaneous Materials

Thank you
for your kind attention

Stan Trout
strout@ieee.org
www.spontaneousmaterials.com

Spontaneous Materials

The Great Misconception:

The grade performance Gap between the alloy block and finished magnet

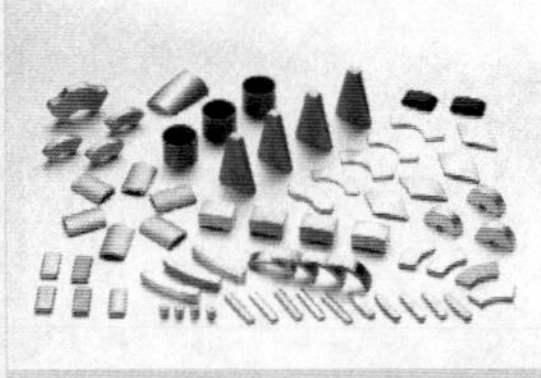

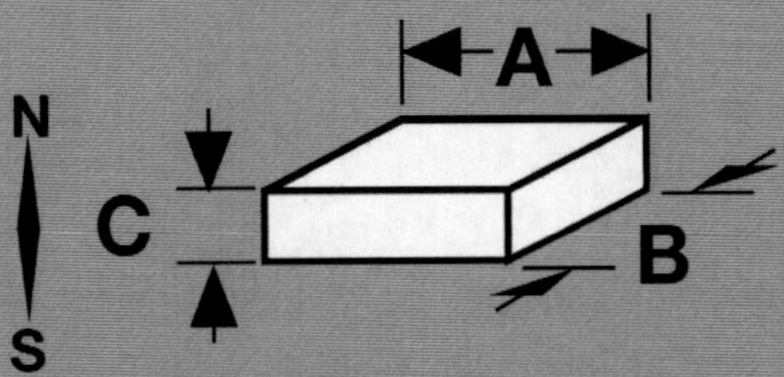

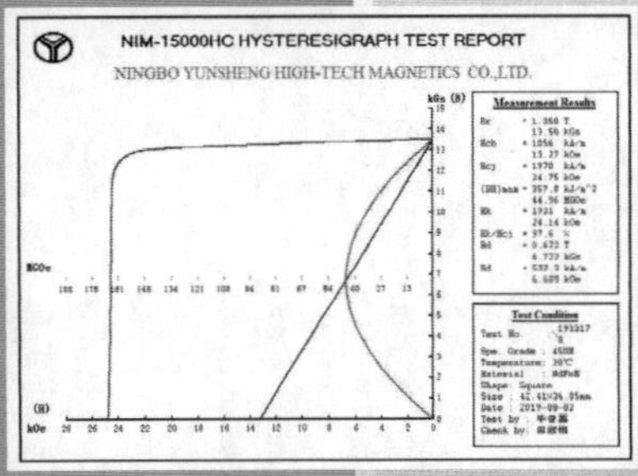

YUNSHENG USA

A Division of Yunsheng Group

Who is Yunsheng?

Founded in 1991; Started Magnet Business in 1995

Listed on Shanghai Stock Exchange in 2000

SSMC License agreement in 2001; Renewed in 2013 with Hitachi Metals

Multiple locations worldwide

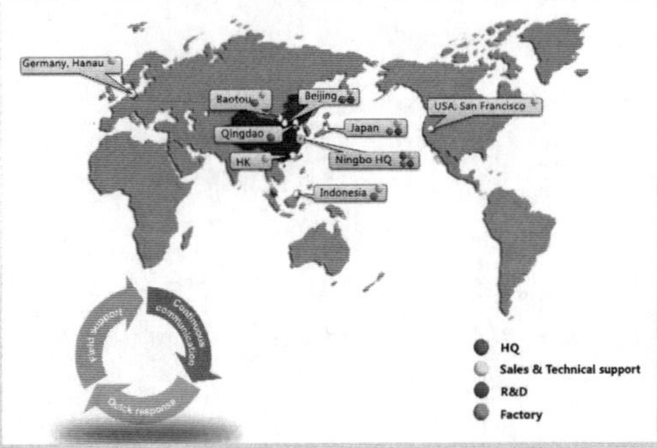

YUNSHENG USA

A Division of Yunsheng Group

What does Yunsheng do?

Yunsheng is the largest neodymium permanent magnets manufacturer in the world.

Block Production capacity of 12000T/Year by end of 2019

- Sintered NdFeB Magnet: 12,000 ton/Year
- Bonded NdFeB Magnet: 400 ton/Year
- Assembly: 10,000,000 sets/Year

YUNSHENG USA

Your Magnetic Possibilities

Park 1-146,000m²
(Ningbo)

*Major Sintered and Bonded Plant before 2014

Park 2-207,000m²
(Ningbo)

Headquarter & Assembly Division

Park 3-140,000m²
(Ningbo)

Sintered Magnet Surface Treatment & Final inspection & Packaging & Logistics

Park 6-28,000m²
(Ningbo)

Sintered Magnet New Surface Treatment Plant (Under Construction 2020)

Park 5-133,000m²
(Baotou)

Sintered Magnet Block manufacturing
Block annual Capacity 6000ton(2019)

Park 4 -80,000m²
(Ningbo)

Sintered Magnet Block manufacturing & machining;
Block annual Capacity 6000ton(2019)

YUNSHENG USA

How do you specify the magnetic performance requirement?

YUNSHENG USA

A Division of Yunsheng Group

Chinese magnet industry standards

Basic Assumption:

Magnet print specifications represent the final magnet grade

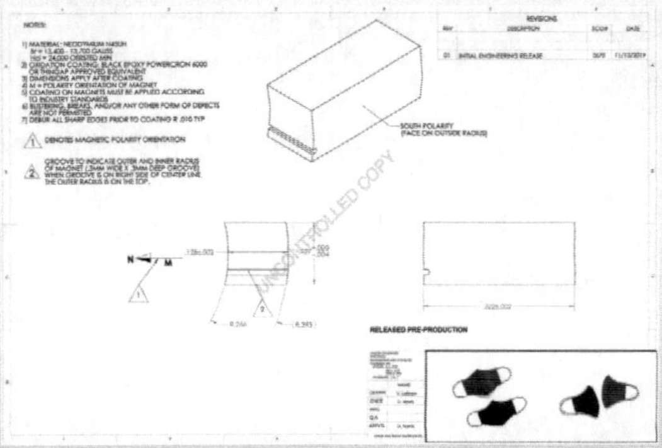

INCORRECT

YUNSHENG USA

Chinese magnet industry standards

Reality : Demag curve reflects **alloy block** grade

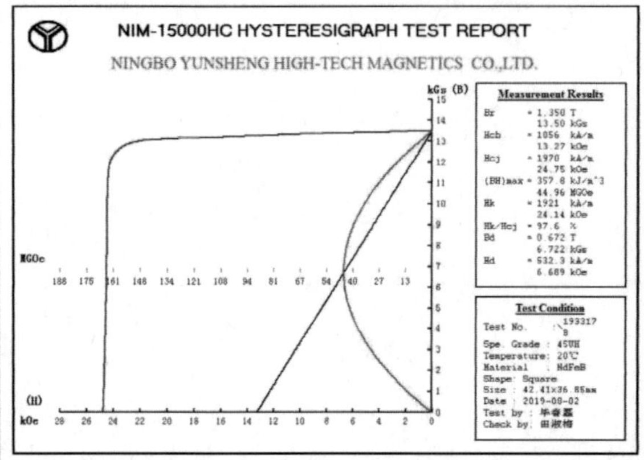

YUNSHENG USA

A Division of Yunsheng Group

Chinese magnet industry standards

- Standard magnetic properties are:
  - Remanence (Br)
  - Intrinsic Coercivity (Hcj or Hci)
  - Coercivity (Hcb)
  - Maximum Energy Product (BHmax)
- Magnetic properties are based on **Alloy Block**
- Demag curves based on **Alloy Block** properties

  Varies from batch to batch ±1.5%
- Alloy block Remanence *degrades* during production process

Magnet Br ≤ Alloy block Br

YUNSHENG USA

A Division of Ningbo Yunsheng

Production Process

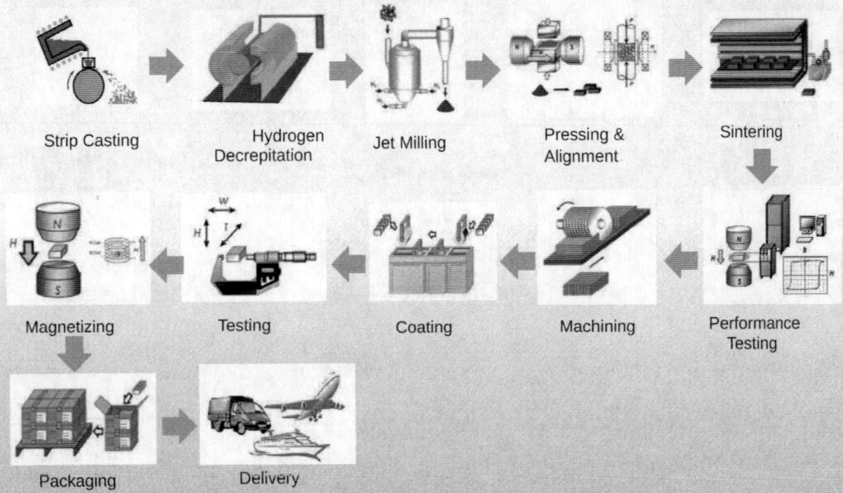

YUNSHENG USA

A Division of Yunsheng Group

Production Variation Factors

Variation introduced during alloy block production

- **Pressing Phase**
  Magnetic alignment of powder
- Sintering Phase
  Location / development of " Hot spots" (eg. Center vs Corner)
- Machining Phase
  Magnetic angle of error during initial "facing" and subsequent cutting
  Geometry of magnet (block vs disc vs ring vs irregular)
  Damage to edges/ corners from chamfering and edge breaks
- Plating Phase
  Etching damage to surface
  Magnetic (minor) interference by plating material

YUNSHENG USA

Pressing Phase:

Variation in

Alignment of dipole

- Anisotropy determined by pre-alignment of magnetic dipoles during Pressing phase
- Alignment fixes material orientation for Machining phase

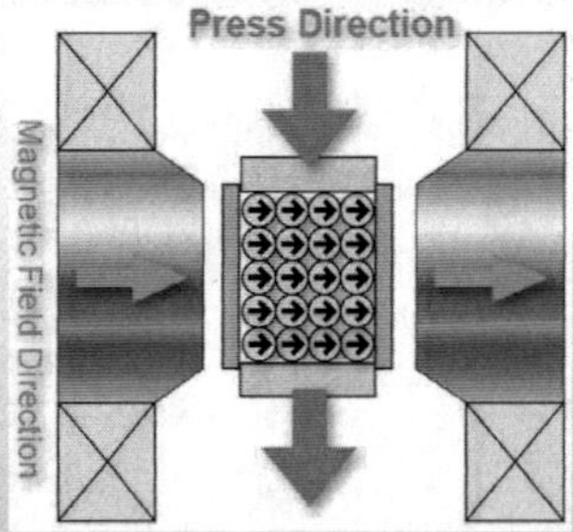

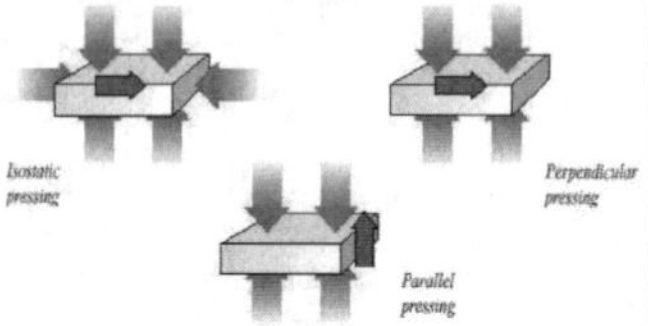

- A variety of pressing arrangements results in variations in dipole alignment

YUNSHENG USA

A Division of Yunsheng Group

Pressing Phase:

Standard distribution of magnetic angle of error

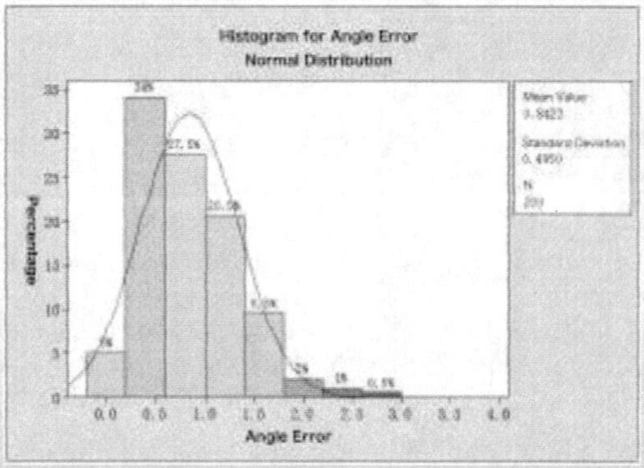

- Relatively high level of accuracy is achievable for *initial* angle of error in green compact
- Normal distribution generally ranges from ±0.5° to ±1.0°

YUNSHENG USA

A Division of Yunsheng Group

Production Variation Factors

Variation introduced during alloy block production

- Pressing Phase
  - *Magnetic alignment of powder*
- **Sintering Phase**
  - *Location / development of " Hot spots" (eg. Center vs Corner)*
- Machining Phase
  - *Magnetic angle of error during initial "facing" and subsequent cutting*
  - *Geometry of magnet (block vs disc vs ring vs irregular)*
  - *Damage to edges/ corners from chamfering and edge breaks*
- Plating Phase
  - *Etching damage to surface*
  - *Magnetic (minor) interference by plating material*

YUNSHENG USA

A Division of Yunsheng Group

Sintering Phase: Location and development of "hot spots"

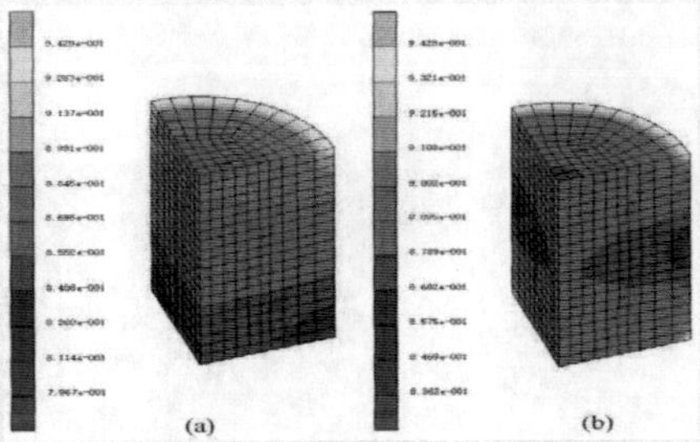

- Relative density and powder mesh size
- Heat-affected zone in sintering oven
- Dwell time in compaction/sintering process

YUNSHENG USA

A Division of Yunsheng Group

Production Variation Factors

Variation introduced during alloy block production

- Pressing Phase
  Magnetic alignment of powder
- Sintering Phase
  Location / development of " Hot spots" (eg. Center vs Corner)
- **Machining Phase**
  Magnetic angle of error during initial "facing" and subsequent cutting
  Geometry of magnet (block vs disc vs ring vs irregular)
  Damage to edges/ corners from chamfering and edge breaks
- Plating Phase
  Etching damage to surface
  Magnetic (minor) interference by plating material

YUNSHENG USA

A Division of Yunsheng Group

Machining Phase: Initial Facing operation

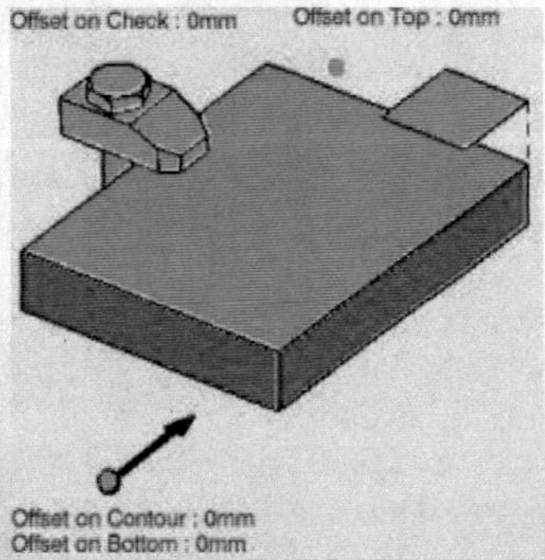

- Initial Facing operation for alloy block **fixes** angle of error
- Subsequent machining steps **increase** angle of error

YUNSHENG USA

A Division of Yunsheng Group

Machining Phase:

Impact of geometry on Br

- Br is a measurement of surface flux
- Curved edges and arcs complicate Return path, *distorting* surface flux patterns and measurable Br
- Irregular geometries create more complicated Return paths, resulting in even more *unpredictable* surface flux patterns

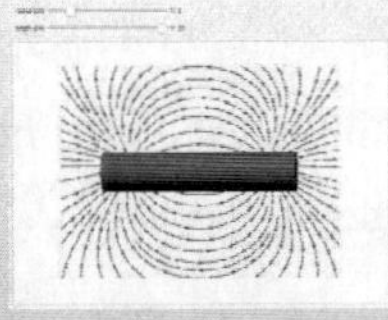

Cylinder magnets

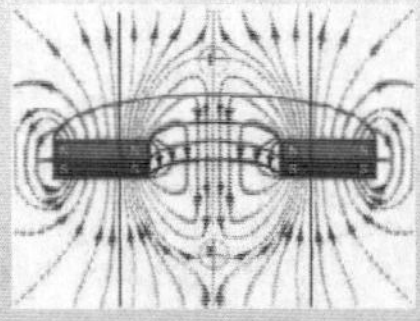

Ring magnets

YUNSHENG USA

A Division of Yunsheng Group

Machining Phase:

Variation due to geometry

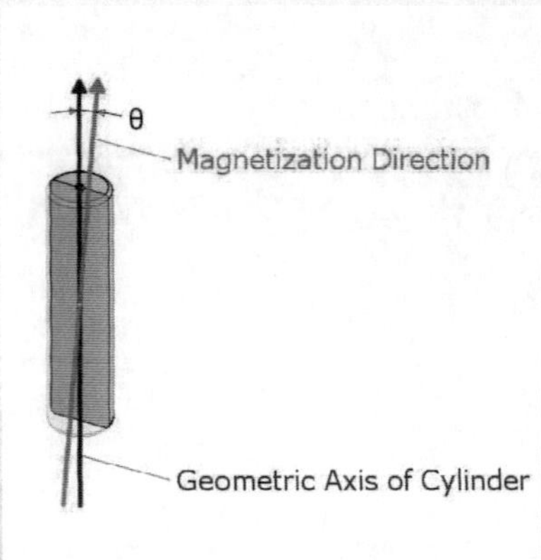

- Factory Guaranteed angle of error in magnet is ±5.0°

YUNSHENG USA

A Division of Yunsheng Group

Inspection: Angle of error measurement

Standard Helmholtz coil instrumentation

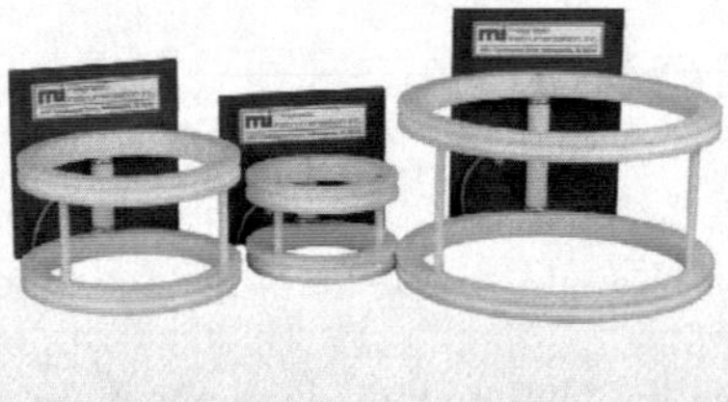

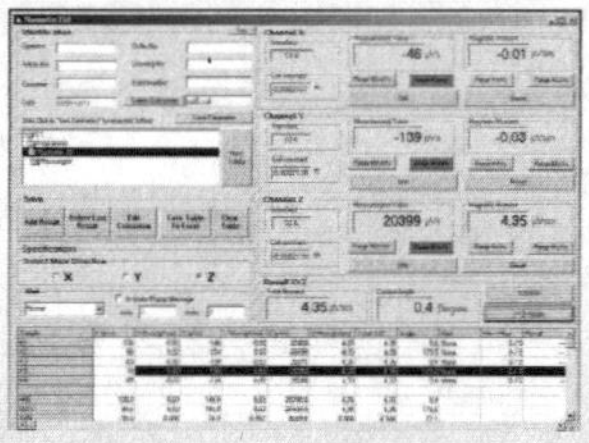

- Off-the-shelf equipment – no fixture used
- Measure flux at X, Y, Z axis in the coil
- Angle of error calculated by software programme

YUNSHENG USA

A Division of Ningbo Yunsheng

Inspection: Angle of error measurement

Yunsheng Customized Helmholtz coil instrumentation for repeatability

- Affix the Magnet into fixture board slot
- Place fixture board into the coil
- Measure flux at X, Y, Z axis in the coil
- Angle of error calculated automatically by software programme

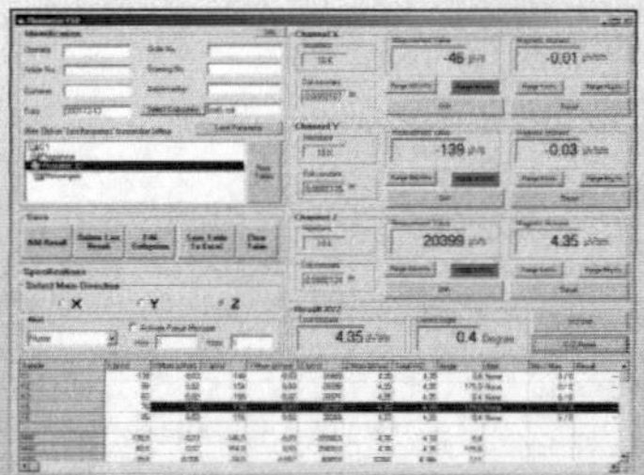

YUNSHENG USA

A Division of Yunsheng Group

Production Variation Factors

Variation introduced during alloy block production

- Pressing Phase
  Magnetic alignment of powder
- Sintering Phase
  Location / development of " Hot spots" (eg. Center vs Corner)
- Machining Phase
  Magnetic angle of error during initial "facing" and subsequent cutting
  Geometry of magnet (block vs disc vs ring vs irregular)
  Damage to edges/ corners from chamfering and edge breaks
- **Plating Phase**
  Etching damage to surface
  Magnetic (minor) interference by plating material

YUNSHENG USA

A Division of Yunsheng Group

Plating phase: substrate damage

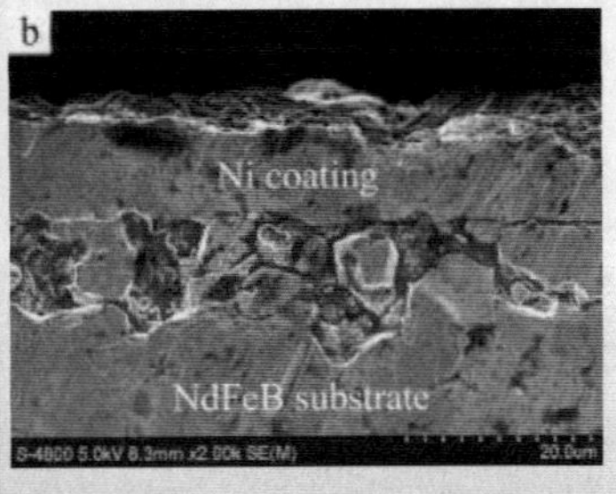

- Preparation of NdFeB substrate for electrolytic Nickel plating involves "etching" solutions
- Etching chemicals clean - but also cause microscopic damage to surface layer, particularly at the magnet edges and corners
- Damage from plating phase can lead to less predictable (distorted) flux Return paths at magnet corners and edges

YUNSHENG USA

A Division of Ningbo Yunsheng

Conclusions: Degradation of Remanence from Alloy block to final magnet

Production factors contribute to degradation of alloy block Br:

- ❖ Magnetic alignment during Pressing phase
- ❖ "Hot spots" within alloy block during Sintering phase
- ❖ Magnetic angle of error during initial "facing" and subsequent machining steps
- ❖ Geometry of magnet (block vs disc vs ring vs irregular)
- ❖ Damage to edges/corners from chamfering and edge breaks
- ❖ Etching damage to surface by Plating phase
- ❖ Magnetic (minor) interference by Plating material

Magnet Br ≤ Alloy block Br

YUNSHENG USA

Conclusions: Grade expectations from Magnet print specifications

Adjust grade Expectation

- Print Br specification should be based on alloy block Br to avoid potential confusion with draftsman or engineers, or Supplier
- Magnetic grade is "reference only" for final magnet
- Reach an agreement with Supplier on acceptable batch to batch variation for alloy block Br

Magnet Br ≤ Alloy block Br

Final magnet Specification is a MUST-HAVE

- Be willing to move up a grade (or two) in alloy block Br
- Reach an agreement with Supplier on acceptable batch to batch variation for alloy block Br needed to achieve final magnet Br
- Reiterate this special requirement clearly with each, new magnet design for Supplier

YUNSHENG USA

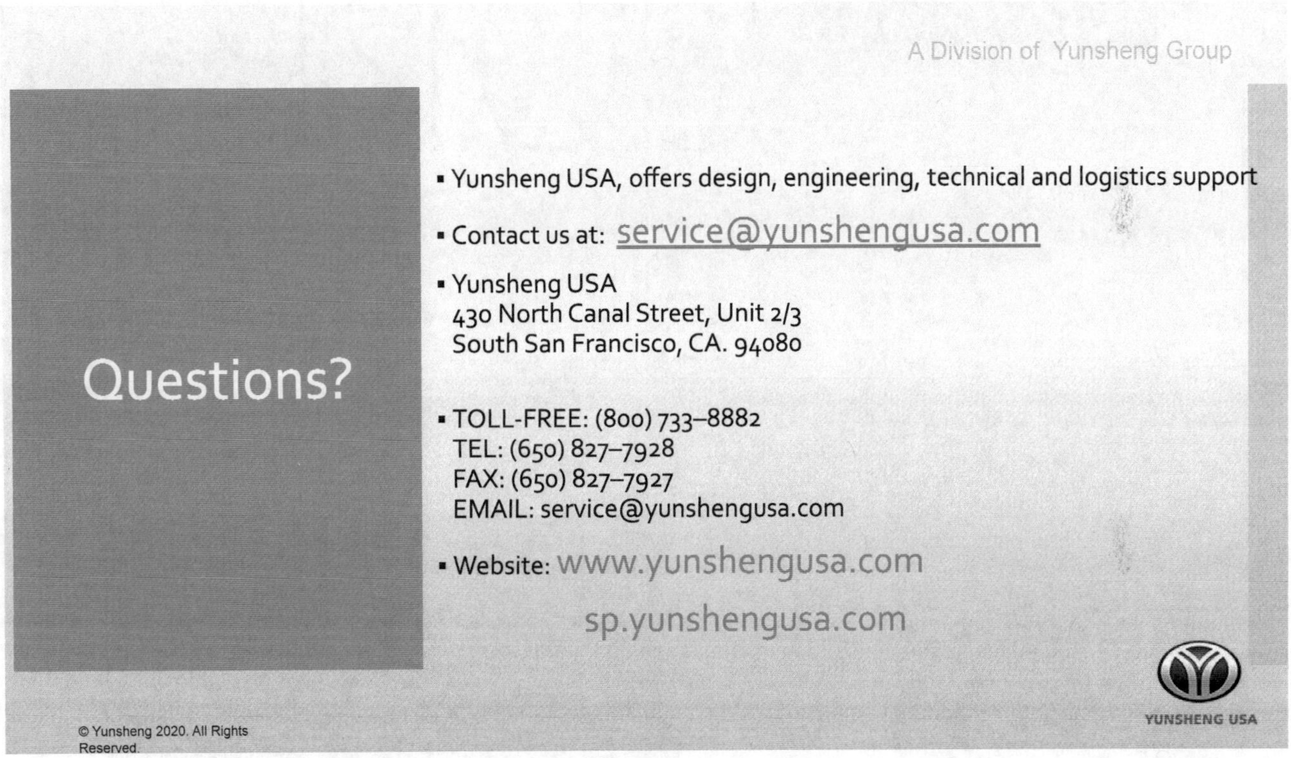
A Division of Yunsheng Group
Questions?
Yunsheng USA, offers design, engineering, technical and logistics support
Contact us at: service@yunshengusa.com
Yunsheng USA
430 North Canal Street, Unit 2/3
South San Francisco, CA. 94080
TOLL-FREE: (800) 733–8882
TEL: (650) 827–7928
FAX: (650) 827–7927
EMAIL: service@yunshengusa.com
Website: www.yunshengusa.com
sp.yunshengusa.com
YUNSHENG USA
© Yunsheng 2020. All Rights Reserved.

MAG™

Magnetic Answers for Market-Driven Needs

CORP

UN & UP

Serving Unmet Needs and Underserved Populations

UN&UP Leverages Non-Dilutive NIH Funding to Develop High-Impact Medical Technologies and Therapies

UN&UP Core Technologies

Magnet-Based Robots for Minimally Invasive Surgery

- Electrophysiology Ablation
- Interventional Neuroradiology
- Interventional Pulmonology
- Minimally invasive Neurosurgery

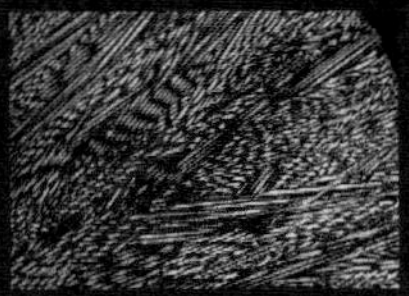

PtCo ingot XRD

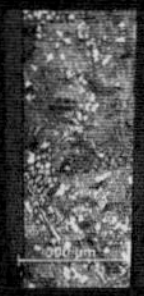

PtCo Wire

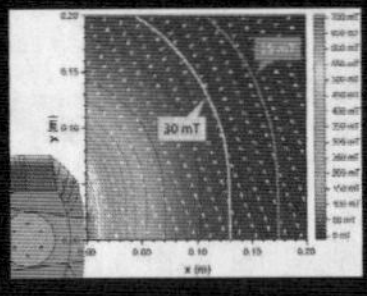

Sextapole magnet

Magnetic Particle Delivery Systems for Clinical Therapy

- Hepatic Carcinoma Embolization
- Neuroprotectant Delivery
- Glioblastoma tumor chemotherapy
- Thrombolytic Delivery for Stroke

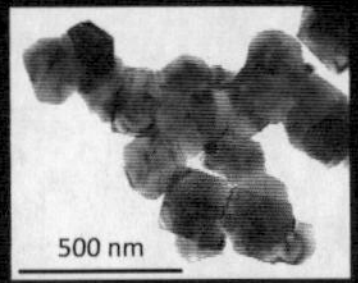

Iron oxide nanoparticle

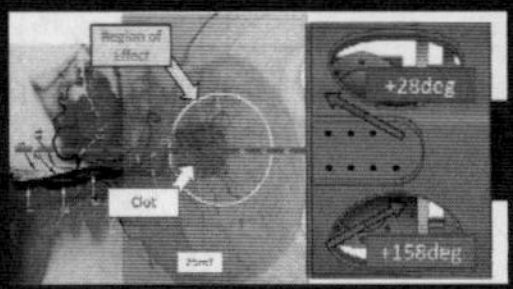

Magnetic collection system

UN&UP – Magnet-Based Robotic Systems

Magnetic Catheter

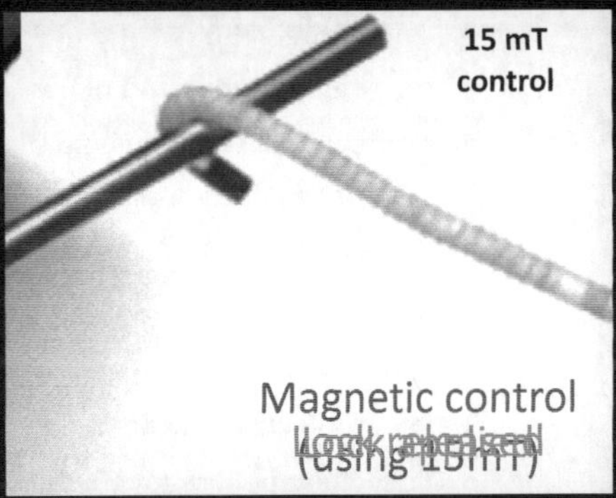

Magnetic Guidewire

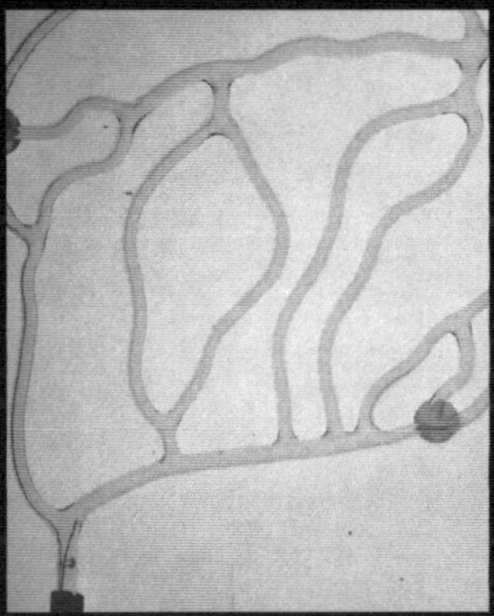

UN&UP – Magnet Particle Delivery Systems

Magnetic Focus

Alternating the magnet's rotation causes nanoparticles to climb the opposite side of vessel

4mm culture tube

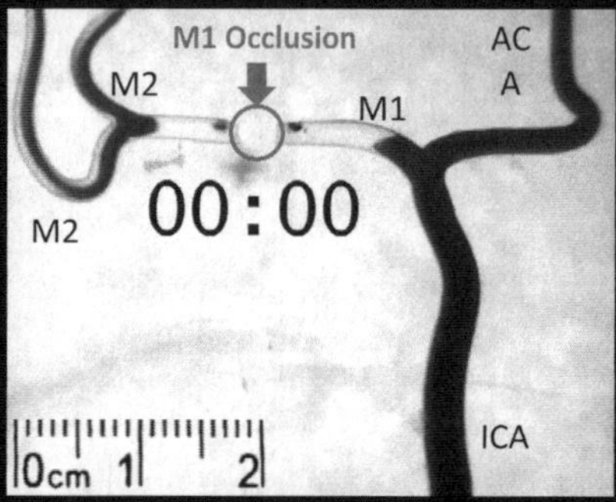

- 5X speed (MIN:SEC shown)
- 60 BPM modeled flow
- Blue dye used as tPA surrogate

Cross Pollination

MAG™ CORP

U & N U P — SERVING UNMET NEEDS AND UNDERSERVED POPULATIONS

LOW MAGNETIC FIELD MEDICAL ROBOTICS

- Electrophysiology
- Interventional neuroradiology
- Interventional Pulmonology
- *Five NIH SBIR awards since 2019 (NINDS, NHLBI)*

MAGNETIC DRUG DELIVERY

- Hepatic Carcinoma
- Neuroprotectants for Brain Trauma
- Chemotherapeutics for Glioblastoma
- *Three NIH SBIR awards since 2019 (NCI, NINDS, NIMHD)*

NATIONAL HIGH
MAGNETIC FIELD LABORATORY

- The **only facility of its kind** in the United States and the **largest** and **highest-powered magnet laboratory in the world.**
- Home to magnets up to two million times stronger than the earth's magnetic field.
- Exists to provide a research environment for U.S. and international scientists who come to make **foundational** discoveries to advance basic science, engineering, and technology in the 21st century.

DOMESTIC INSTITUTIONS	INTERNATIONAL INSTITUTIONS
133 UNIVERSITIES	113 UNIVERSITIES
20 GOVERNMENT LABS	25 GOVERNMENT LABS
20 INDUSTRY	13 INDUSTRY

The Problem Solving Gap

THE MAGLAB EXISTS TO SOLVE PROBLEMS OF THE **FUTURE**

WHO CAN SOLVE THE PROBLEMS OF **TODAY?**

The world knows the MagLab is the crown jewel of magnet science.
What is still needed is a way to tap into this expertise and infrastructure to solve the problems industry faces now.

This is why we created MagCorp.

We put the MagLab to work for you.

MAG™ CORP

Who Do I Call?

How Do I Start?

Do They Have What I Need?

What Will The Deal Look Like?

How Long Will This Take?

How Much Will This Cost?

Where Will This Be Done?

...?

Consulting

- Finding The Right Expert
- Negotiation
- How To Pay Expert

Sponsored Research

- New contact at Sponsored Research
- Negotiations
- Pay Sponsored Research

Product Development

- New Contract & Licensing Manager
- Negotiate With Manager
- Vendor Approval

We Have The Answers.

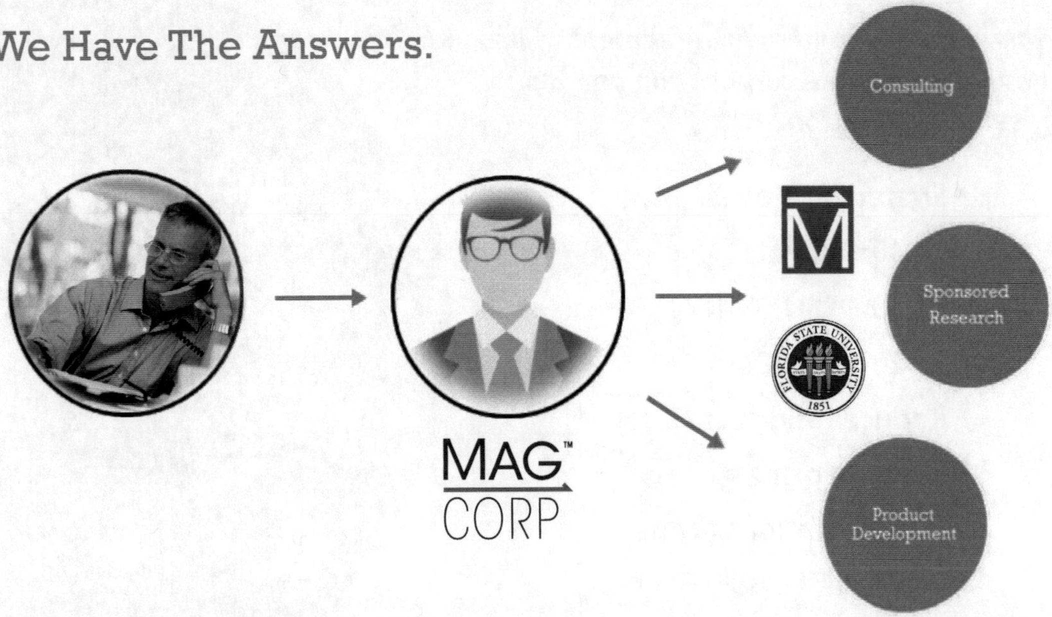

We didn't make this up.

We listened.

1) Anxiety over IP & indirect cost negotiation process
2) Relatively slow communication & agreement execution timelines
3) Everyone talks science, and nobody talks business

No advocate has ever existed for facilitating communication.

This is the first and only agreement of its kind between a private corporation and a public research institution.

- **Shared IP Model**
  - 40% Client
  - 40% FSU/NHMFL
  - 20% MagCorp
- **Reduced Indirect Costs**
- **Master Format**
  - Expectations are set and projects begin faster

MagCorp has executed a Master Services Agreement with Florida State University.

The MagCorp Stepwise Engagement Model

- ❑ Begin engagements with more easily fundable, low risk feasibility studies and proof of concepts
- ❑ Progress to bigger investments in prototyping and commercialization on validated projects

- Lowers the risk to all parties
- Swift executing agreements and quickly responsive
- We speak the same language
- Complete operational control & flexibility
- Platform for tech transfer of SBIR/STTR phase 3 projects

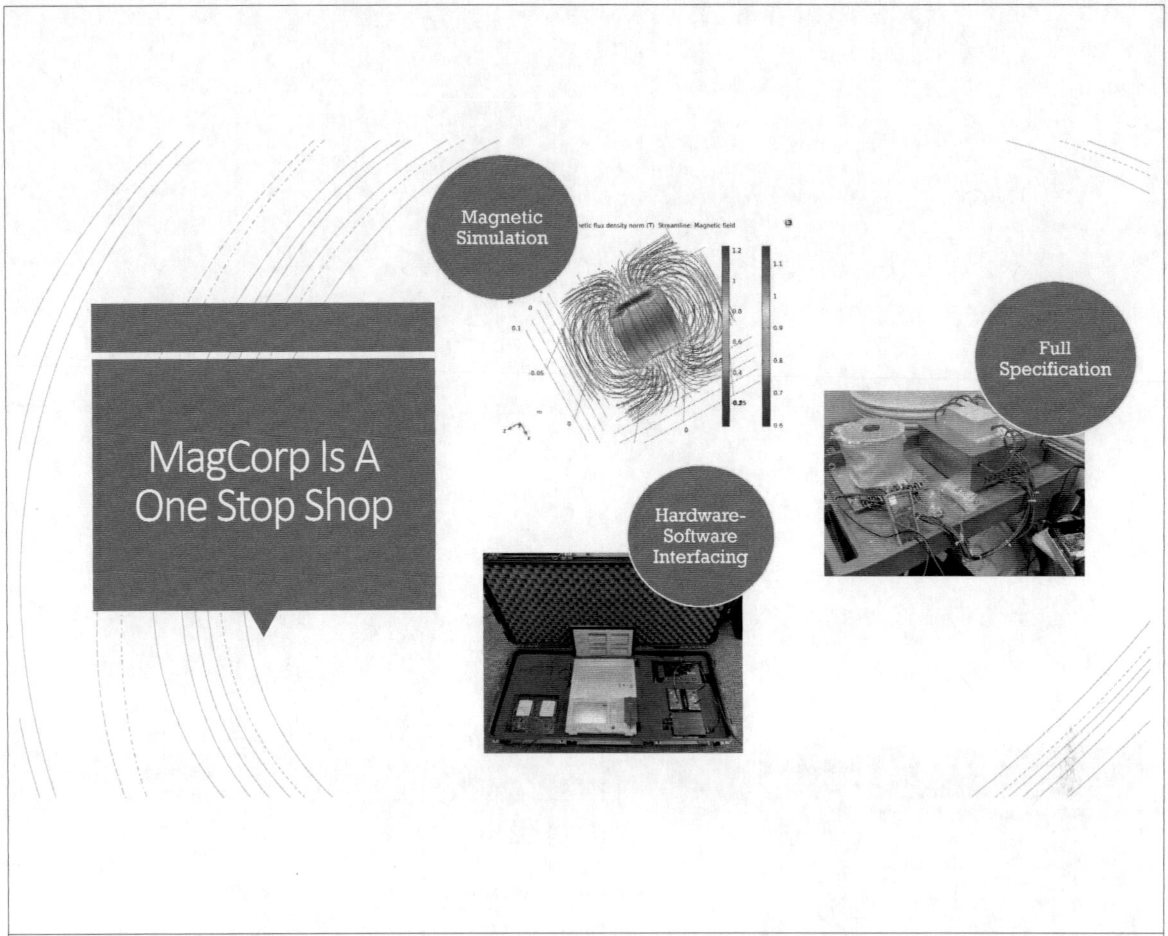

MagCorp Is A One Stop Shop
Magnetic Simulation
Full Specification
Hardware-Software Interfacing

What Does A MagCorp Project Look Like?
Just some examples:
Military
Manufacturing
Consumer Products
Energy Production
Communications and Sensors
MedTech
Recycling
Textile Products

Contact Us

- Dr. Jeffrey Whalen, Director
  - Jeffrey@magneticscorp.com
  - (850) 321-5540
- Lezlee Richerson, General Manager
  - Lezlee@magneticscorp.com
  - (850) 228-8009
- Abby Queale, CEO
  - abby@magneticscorp.com
  - (850) 296-2553

AUTHOR INDEX